fingers in the sparkle jar

fingers in the Sparkle jar

Lessons in Life and Death

chris packham

EBURY PRESS

1 3 5 7 9 10 8 6 4 2

Ebury Press, an imprint of Ebury Publishing
20 Vauxhall Bridge Road
London SW1V 2SA

Ebury Press is part of the Penguin Random House group of companies
whose addresses can be found at global.penguinrandomhouse.com

First published by Ebury Press in 2016

This book is a work of non-fiction based on the life of the author.
While a handful of names of people and places have been changed
solely to protect the privacy of others, the vast majority remain
correctly identified.

www.eburypublishing.co.uk

A CIP catalogue record for this book is available from the British Library

ISBN 9781785033483

Printed and bound by Clays Ltd, St Ives PLC

Contents

I

Brand New Savage

The Collector

July 1966

'I'M SORRY, I haven't got change of a ladybird.'

The ice-cream man had opened the matchbox expecting a sixpence but instead found a six-spotted beetle that was now scuttling manically over his counter, defiantly refusing reinterment in its crisp little cell despite repeated repositioning. He gently pressed his cupped palm down on the fugitive and as it squeezed free of his fingers managed to flick it back into the box. Green juice specked his nail. He tutted, wiped it on his trousers and stretched over to hand the doomed specimen back to the beetle-keeper just as a girl he knew as Anna simultaneously thrust a thick wodge of bubblegum cards at the little mush and demanded, 'Why don't you let it go?'

The boy ignored her, shook the box next to his ear, fanned through his worn wad of Batman and Tarzan cards and then wrestled them into the bulging pocket of his shorts.

'You wouldn't like it in there,' she snapped before she looked up to Mr Whippy and said, 'He's had it in there for days and he never lets them go until they die. I'd like a Sky Ray.'

Yellow chin lights speckled her freckles, radiating from three daisy chains, her morning's work. Ghostly grey eyes and pig pink cheeks cuddling close to the grubbiest teddy bear. She had sunblown blonde play-curled hair that flopped in tattered curtains over her milky brow and tickled those tickly arcs of soft skin beneath her eyes. The boy was skewy-fringed and silent, gazing at the pavement in mandatory scuffed sandals, and both wore too-tight over-washed T-shirts, hers with a smiling fox's face, his with fresh pearls of Airfix cement constellating its front.

'Please. I'd like a Sky Ray *please*,' she repeated to be sure her manners would be recognised.

The suns flaring from the side of the van lit the scruffy waifs up like camera flash as they squinted hard at the pictures of lollies in awe of Strawberry Splices and Orange Sparkles. Her fist unfurled into a cup of sweaty pennies that would smell of bitter money till bedtime and when she came to lick the melted lemon as it dribbled down the stick that metallic taint would bite back and make her wince and spit and her big sister would scoff and giggle.

The bloke picked out six coins and handed her the Sky Ray, which she unwrapped fiercely, leaving strips of paper glued to its frosted sides. She sucked at its cherry red tip and after a pleasurable pause poked the boy, who flinched and magically produced a thrupenny bit for his thrupenny ice-lolly. As they left a gang of panting nippers with rattling trolleys charged up. These dad-made pavement racers rolled on rusty wheels scavenged from dumped prams and pushchairs and were clad in garish strips of threadbare carpets and daubed with the sticky dregs of their parents' house paint pots. They steered them roughly with their rope loops into the wall and all at once told him there'd been a massive crash in the woods over by the council flats yesterday and that Axell had broken his arm and been taken to the General by his mum. On the bus. He was their hero, king of the juniors. They brandished their scabs, which they squeezed to get fresh blood, and without a shilling between them stormed off to kick a ball against a wall until teatime and *Crackerjack*.

With the kids gone the ice-cream man had a quick fag and picked himself out a Woppa. When he slid the freezer shut he saw the 'ladybird boy' standing staring at him from the lawn beyond the wall. From each hand dangled a shiny jam jar and it was painfully obvious that he had returned with these trophies because he wanted to share them. The man checked his watch, then flicked his head to beckon him over. The boy snaked across the grass to the gate, put both jars down, clicked the latch, moved them outside, setting each gently on the tarmac before dragging the hinge-less flaking panel shut behind him. He then carefully laid out the string handles and picked up the pots synchronously and slowly so they didn't swing. These were sacred things.

The man leaned across the counter and out of the window and the boy offered him one. Through a mist of condensation he could see a muddy base, three or four large pebbles, several sprigs of wilted greenery and a pink plastic dinosaur. It was standing upright, snarling, and just visible beneath it was another pale blue prehistoric form lying on its side. The prey. He scrutinised it; everything was very precisely arranged, a perfect diorama modelled on some encyclopaedia's illustration of the world one million years ago. He rotated it carefully, judged he had spent just enough time in rapt appreciation and handed it back.

The second jar held a similarly contrived miniaturised scene but had no discernible ancient reptiles. Its sides were also dripping wet but its lid was roughly ventilated and so presumably housed something living. But the amiable geezer couldn't be bothered to look that hard. He passed it down and the boy frowned as he adjusted the handle to ensure symmetry and then looked back at him, his eyes slit tight against the hot bright light. He should say something, but what? What do you say to a weird kid with dinosaurs in jam jars who never speaks, who only ever points, who buys your cheapest ice-lollies and seems to think that bartering with various bugs is a viable currency for exchange? So he nodded, a nod which he hoped would signify approval, and grunted, 'What happened to your lolly?'

The kid put down the jars with robotic precision and then pulled the stick out of his pocket and held it up. He then reached into his other pocket and withdrew a bundle of about twenty more bound to a small spoon with a bent neck. As he drove off Mr Whippy nodded again and when he checked his mirror at the corner he could see the kid still kneeling on the pavement. Peering into the pot with nothing in it.

Back in the garden the boy carefully unscrewed the lid. The jar belched a bitter breath and peering into the sweaty cell he found this afternoon's ladybirds running unnaturally fast, baked into a panic, scuttling and slipping on the wet sides. He set the lid and jar at the dead centre of a paving stone then rested on his elbows to watch them all escape.

He could see their flat feet on the grey glass, winding and failing, black-bellied, oval, with their feelers tapping feverously, struggling to find fresh air. The first beetle made it to the lip and at once began to circle the rim, pausing to change direction and crack open its ruddy back, trapping a twist of orange tissue in a momentary tail. Round and round, another appeared and then a third, a crowd, busy bumping, he knew what they wanted. He shuffled across to the unkempt edge of the lawn and pulled a stalk of grass, bit off the limp base and then the flower spike to leave a long straight straw that he placed in the jar leaning against the edge. At once the gyrating carroty beads climbed it and at its pinnacle took flight and wafted away. Finally, as the last prisoner reached the gleaming parapet and trundled to the base of the launch tower, he dragged the jar closer and watched the mini machine's six legs organise the ascent. He traced its line up the stem and placed his finger at the tip so the insect crawled onto it.

As it turned he swivelled his wrist to face its front. Two white patches like eyes above its tiny head and waving, it stood up flailing legs and falling back, then shuffled and settled, and he knew it was time. The ripe little ruby split, its concealed wings unfurled, it lifted, hovered, twinkled and sunspun up, glistening for a second before vanishing into his piece of sky.

He shook the fetid salad out of the jar onto the lawn and pawed at several blackened grains, which had expired before the great escape, crushing one on the tip of his finger. It smelled sharp and strong and stained his skin brown despite a feverish wiping on his shirt. He wondered at how quickly they died and tried to imagine how many died every day in the whole world. It was sad that something so beautiful, so perfect had to die, it seemed unfair and wasteful and was difficult to understand. Things were either dead or alive, they were alive right up to the point they were dead but there didn't seem to be anything in between. They could be a little bit alive but they were always completely dead. He didn't know any dead people but one day he would, he knew some old people, they would die, maybe soon, maybe even before he had his next ice cream.

He wondered how many of the old people he'd seen but didn't know had already died. He would die too, one day. But he didn't want to. All this thinking about dying made his chest ache, he was nearly crying. There was a dead ladybird stuck in the bottom of the jar. He shook it hard to get it out but it wouldn't budge until he used the grass straw to dislodge it and flick it away. He didn't like it dead.

He sucked the mark on his shirt and scrubbed his fingertip again and then straightened the T. rex that had fallen over the Brontosaur in his other jar. Then he took out his matchbox and climbed onto the wall alongside Mrs Greenwood's ladybird bush. It was teeming with blood-red beetles and soon his fingers were stained by their little bleedings and he was shaking the box to listen to the many captives rattling inside. He'd put them in the jar later but now he jumped down and lay on the dusty lawn. Clicking gurgled squeaks rattled and rhythmically wheezed from the bird on the chimney pot whose glossy feathers flashed bits of rainbow from its spiky crest. He lay back and whistled, the bird spluttered on, he wet his lips and whistled again adding a flourish and the bird whistled back mimicking his notes, he waited, the bird rambled through its repetitive repertoire, then he whistled again and the bird replied. The duet went on until the mimic vanished and then he whistled and answered himself, stroking the polka dotting of daisies with sweeping arcs of his arms, in synchrony and symmetry. He swam in his paradise, his heaven of a million living things.

The Suburb

August 1966

Bright and polished, stacked up like a giant cake iced in layers of brilliant white, sky blue and rich buttery yellow. Chrome quarter bumpers at the back flashed pricks of silverlight and crowning the cab above the windscreen were two freshly whipped cones, each stabbed by a matt chocolate flake. The sun blazed through these plastic lamps and the van's roof and fluoresced softly on the pages as he pawed over the morning's news. Myra Hindley was glowering at him, black-eyed, blonde but bloke-ish and so obviously ruddy evil. He turned it over, she was too creepy.

Predictably there were the boring Beatles and the England team, Moore brandishing the trophy, Stiles's toothless grin and Charlton looking like his old man. Absolutely bloody marvellous he thought as he tossed the paper on his seat and leaned forward to turn on the radio. The Kinks had it right, he was indeed lazing on a sunny afternoon, in the summertime. He hummed it, filling in between the crackles until the tranny warmed up and settled down.

At the weekend, the kerbs would be dotted with Cortinas and Victors, Minxes and Imps but this afternoon the sideways were the wickets and Wembleys for England's fantasising finest. Everywhere there were kids telling tales and blowing bubbles and as he'd steered his way through the roads, crescents and closes, diligently checking his list of stops, his cash tray quickly filled with their pocket money. He passed a row of overcoated old codgers at a bus stop, avoided stray mongrels skipping over walls and dodged abandoned tricycles and bicycles. He watched retired men mowing tiny lawns, their wives hoisting high lines of washing, he heard the ancient strains of grannies and nannies wringing old mangles and saw a bright new football rolling lost in a grassy gutter whilst the goal scorer was at home flob-a-dobbing with the Flowerpot Men. As the last notes of 'Greensleeves' crackled from his loudspeaker he pulled up where Cornwall Road met Cleveland Road at four thirty on the nose.

As he stood in anticipation, arms spread on the counter, he looked up the hill at this stretch of twenty or thirty pebble-dashed houses and smiled. They were clean, tidy, the gardens in front of him were hoed and manicured, filled with ordered greenery, each gate had been carefully closed and painted in colours to match the windows and doors, many of which were neatly two toned. Purple and lilac, dark and pale blue and one in striking yellow and black, the only one with wrought-iron gates. Butterflies jostled over flower beds, a blackbird listened to the lawn, a large tree was laden with clusters of flaming berries, there were house names and numbers and birdsong, a portrait of a perfect suburbia.

It was also an ant day. The winged masses were rising from a drain immediately in front of the van, a steaming fountain of glossy scales erupting into the hot afternoon sky and this geyser of insects shimmering there in a whorl of twinkling wings showed no sign of slowing up. There must have been millions and millions of them leaving their dirty womb, swarming over a broken blue eggshell and into the sky. How on earth could they all have grown to do this? What tonnage of sweet things had been stolen in how many tens of thousands of forays by the manic workers through the minuscule chasms of this neighbourhood? He played the chimes and still they rose, so hypnotically that his first customer arrived unnoticed. The 'ladybird boy' from last week.

But in an instant he was brushed aside by a soldier, an astronaut, a cowboy and an Indian, a Liverpool striker and two ballerinas. Assorted snot and the smell of fired caps, a few new plimsolls, they spent their pennies and sped away, their dirty little hands catching hundreds and thousands, tiny tongues licking orange, chocolate and blackcurrant drips from pinkening palms, running with stiff arms clenching their soggy cones, staining their dresses with dribbles and their smiles with cherryade, disappearing into the sunniest summers they'd ever have.

There was a birthday party at number seven. He watched them chasing themselves, hysterically happy, and then turned to find the boy had lingered. He was holding a crocodile. He was cradling the stuffed reptile as if it was alive and not the tattered and broken, sparsely scaled ruin that had been abused by a taxidermist years ago. There was no danger of a crocodile being proffered in exchange for an ice cream though; this was a prized possession. Nevertheless it was offered up and he took it. Its earhole was unusual and prominent, its teeth clean, white and needle-pointed, cotton wool was visible behind its glass eyes, its body was bloated and its tail disintegrating around a rusted twist of wire. It was pretty horrible. He handed it back down. A flaky scale lay on the counter and he pinched it up and reached through the hatch to push it into the boy's grateful palm.

'Do you want anything?' He gestured at the lolly pictures.

Without looking up, the boy shook his head.

'Have you got any money?'

He nodded and withdrew a coppery tinkle of coins from his pocket. And then he walked, not ran, off, cupping the scale, which was worth more than all the pennies pocket money could provide. He disappeared through the blue gate, head down, in love with his dead pet.

It was peaceful. The *Animal Magic* theme plinked around in the distance, two poodles peeked through the side gate of number eight and a spangled starling pitched and scoped the pavement for crumbs. He threw out some broken wafers and was away before the squawking started.

The boy listened to the noisy party across the road then trod the stairs centrally, opened and closed his bedroom door with his left foot and placed the juvenile caiman – it was not a crocodile – reverentially on his chest of drawers. Then he sat in a preordained spot, his toes touching the dragons he'd drawn on the lino and from where his view of the reptile's mutilations were minimised. He studied it intensely, rocking his head imperceptibly to improve, with ample imagination, the idea that this withered fragment of worshipped wildlife was actually still alive.

Outside sparrows argued, the sun flooded over his wall's mishmash of creased maps and corner-less posters of dinosaurs, Daleks, Spitfires and the solar system. The TV was on downstairs but he didn't care. It was half-time in a Subbuteo match between Liverpool and Everton and he played both teams, it was twelve all. His games were always draws because he played each team until they scored irrespective of the rules of Association Football, a copy of which his father had bought him. Everton only had eight outfield players, the others having been painfully knelt upon. Repeated gluing had reduced them to blobby dwarves and thus imperfect, thus unacceptable and thus euthanised, banished to the battered and collapsed green box beneath the bed. Table football, played on the floor because they didn't have a table big enough for the baize pitch, was okay but it wasn't as good as animals. He turned to his windowsill menagerie, his jam jars, neatly lined up and gently roasting their inhabitants in soft afternoon light.

The cinnabar caterpillars that had trundled so frenetically in their exotic ochre and black stripes whilst annihilating fronds of smelly ragwort had been emulsified and remained only as a brown soup corrupting the bottom of the jar. The last two-legged tadpole, which had bravely outlived so many hundreds of its spawn fellows since March, was now struggling through a mat of choking algae to desperately gulp for warm air. This year metamorphosis would be understood only from books, not witnessed in nature. A coiled bronze bangle gleamed, a tailless slow-worm, too heavily petted, perhaps too long confined. Fifteen minnows, a tortoiseshell butterfly, three male smooth newts and too many garden snails were already 'gone' and the worm jar was ominously still. However, a riot of glimmering life was exploding in the central Robinsons repository: between three and five hundred newly emerged queen ants were circling with a furious urge to meet and mate with the males who were somewhere outside, rising into the cooling sky, feeding screaming squadrons of happy swifts. These celibate spinsters were rapidly losing their wings, the glass base was already gilded with a fragile skin of golden tiles and the lumpy virgins were tumbling through the kaleidoscope of fractals with diminishing vigour. By morning they too would have all but expired, the last old maids just twitching before his next safari would set off to nonchalantly and excitedly replace them.

Glitterlight sparkled through the dancing canopy and lime-lit the compacted soil with a jigsaw of chasing patterns, swishing and mixing as his eyes chased them trying to find regularity, snatching spots and smudges that almost returned as the branches bounced and shade fell for a cloud-bound minute.

He waited, his knees tingling as the pins and needles sharpened, but he wasn't allowed to feel any pain, he must see the patchwork woven again to match the mind map he'd made. He must pitch his template against chaos and critically identify motifs and ornaments of stability, predictability. And so the sun shot a shard of light, the leafscape formed and for a sub-second the soft patches and shadows projected on the smooth path conformed with a precise familiarity. Then he was done – it was measured, it had been essentially controlled.

He rolled over and straightened his bloodless legs, his cheek on the warm earth, the grass soft on his face, stripes of fuzzy green through which he peeped with a squeezed eye and tingling toes.

The vast savannah stretched away until it melded with a rising bank of darkness, glazing the middle distance ... a lake, capturing a bright line of sky that lit the folded reeds and lilies and all the tangle that tumbled from its shores, flickering as rings rang out from the tickles of distant tiny things that twitched on its silver surface and fizzled in peppery swarms on the other side of the garden.

He was lying on a tablet of riches, his wilderness explored: he knew the plains, the forests, the canyons intimately and where all its life lived and hid, the boulders that covered the scaly caverns of woodlice, where quick twisty centipedes were shiny and soft beneath his fingers, the lovely bark where tiny specks of crimson ran and stained those fingers dead red, the corners where secret spiders stood motionless on their soft handkerchiefs and the lake, pool, baby bath, a muddy cradle in which many miracles swam.

On its banks he'd dip spoons for tadpoles and the twitching larvae of mosquitoes. They'd ziggle down in droves as the steel broke the surface and bent all the lines, then they'd relax their fear and drift slowly up, easier to scoop up and transfer to a saucer where against the white he could just see their eyes and bristles and snorkels. Others were fatter, like comma-shaped bogeys, and then there were the un-wettable rafts of eggs that stuck to his fingers or clung to the rim of his dish.

In the stinky sauce that smoked in whorls from the bottom

when he reached down to ransack the leaves there were fierce things with jaws that scissored as he squeezed them, with bulging eyes and robot bodies, creatures that sat still and then picked their way cautiously on legs that appeared from nowhere.

There were maggots with long tube noses, hard tear-dropped bugs that flicked rapidly backwards and stabbed him when he grasped them tightly in his fist as they tried to flee into the tussocks he'd submerged to build a swamp at one end of the oasis. Wasps drank, newts gulped, skaters skidded ... everything was new, everything needed knowing.

The Pet Shop

August 1966

SHE SLID LAST night's whisky glass across her dressing table and tipped the newspaper off her stool. *The Sound of Music* was showing at the ABC again. It was always on, either that or *Dr Zhivago*. Before he'd slammed the door her husband had bellowed from the kitchen that he wanted to see *The Blue Max* at the Classic, a scruffy little place on the high street that was small, smoky and normally ran saucy films. The trailer had been full of old war planes and of equal appeal to him ... that woman.

She'd go, of course, but wished it wasn't 'his' Saturday night, then she could beg him to take her to the Atherley to see *Born Free* again. He'd watched it with her the first time and made a pretty poor job of pretending to have enjoyed it. Virginia McKenna was no Ursula Andress. It hadn't helped that he'd scraped a line of red paint from the side of his precious Jaguar on the way out of the car park, flown into a furious mood and sulked for a week as a result. It was silly but this accident had completely coloured his take on the film so nothing was going to compel him to go again. There was little point in asking, it would only spark a row and there were plenty of those.

She'd loved the lions, the cubs were so sweet and the scene at the end where Elsa remembers the Adamsons and comes back with her own litter had made her cry both times she'd seen it. A couple of years ago they'd had a pair of three-month-old young lions in the shop, just for a day and a night. She had been so excited, they were irresistible, she hadn't been able to leave them alone, stuffing them with milk until their mauve bellies bloated and they fell into a fidgety slumber on her lap. There were some photographs but they all showed the poor little mites' bald necks and backs, which looked awful and had cost her dearly when the buyer knocked her down supposing they were seriously ill. And although it was ridiculous she so wanted to go to Africa, the space, the animals, she'd stand in her shop looking out onto the dreary forecourt, with its dripping hutches and kennels, and dream of a safari, imagining she was Joy, jumping from a Land Rover, rescuing orphaned cubs, calves and chicks. She drained the dregs of the Scotch, which tasted of cigarettes.

Jesus, Viet-bloody-nam, would they ever stop talking about it? It was either that or this week the papers had been full of those poor dead policemen. She leaned back, seized the wireless dial and twisted it to find some music; snatches of the Ike and Tina quivered and faded under the chirps and whoops and when she centred the red line in the prescribed spot the monotonous 'la la la' of 'Yellow Submarine' started up. Christ, they'd been playing it non-stop on that station, non-stop.

Facing the mirror she found herself frowning but had to relax her face to arch her eyebrows; the mascara was still thick from last night and so was the powder. She drew on the lines and turned from side to side to check them, glanced at the clock, huffed, and furiously backcombed her bleached hair into a balloon of fluffy gold. Breathe in, beige polyester slacks, C&A, heeled sandals, a gold charm bracelet and hooped earrings; she sat back and plumped up her breasts, eat your heart out Ursula.

The Broadway was still quiet as she shook the sticky door open, tinkling the bell. Jackie was knelt sweeping out a cage whilst alongside her a kitten sat idly wasting its moment of freedom and Jimmy was craning on tiptoe over one of the aquaria at the back, the electric green weed illuminating his concentration, tetras flashing over his specs. The thick, the musty, rabbity, papery smell of the shop instantly suffocated her generous bouquet of Guerlain. She left the door ajar, tickled the sulphur chest of Pirate the parrot, whose one dry white eye winked reluctantly, and edged into the space behind the till, careful not to stub her toes and spoil the fresh coral polish.

Picking up a dried seahorse that had fallen from an overloaded bowl on the counter, she let out a Saturday sigh. She dropped the fish back on the pile and it slipped off and rattled, a curious crust of ridges and spines winding into a spiralled tail, nice to touch. This was the shop's busiest day and the weather was going to be nice so there'd be plenty of kids gawping into cages and hutches wishing that they could get a parrot to go with their budgie, a chinchilla instead of a hamster, a python to eat their sister.

He'd be there, the boy with the dad in the shirt and tie who wandered uncomfortably around, dipping into his thick stack of library books whilst his son crouched or climbed to peer into every corner, studying each creature so, so seriously. The kid never smiled but came every week, and stayed at least an hour before the bloke nodded at him, politely thanked her, handed him back his own pile of books and then ushered him out.

He'd meticulously examine all her stock but she could tell he wasn't particularly interested in the mammals, except the fruit bats, nor the birds or fish. It was the reptiles that he liked best and his staring contests with the resident spectacled caiman had become a weekly source of amusement for the staff. He'd stand fixated with his nose locked about six inches from the tank whilst the static croc floated, dispassionately returning his gaze through the vertical split in its golden filigreed eye.

Last week he'd bought another seahorse. She held up the spilled one and studied its horsey head with its funny little snout and then carefully positioned it back on top of the tangle in the bowl. As she moved her hand away it toppled out again so she flicked it into the bin as Jimmy told her that five more of the tortoises had died.

The Mouse

September 1966

THE PREDATORY DINOSAUR stood radiating menace over the scarred patina of the ancient desk and defined the only true purpose for plasticine in 1966. As I rotated it clockwise to check its form was as perfect as my five-year-old fingers could fashion, I sneaked a glance leftwards. Karen Harris had made a snake. It looked as if it had been feasting on bowling balls and then been run over by my dad's Ford Anglia. It was ugly but it was, I supposed, at least meant to be a reptile.

It was my first day at school. True to my mother's habit I was late, last to arrive at the red brick Victorian infants where years before my father had crouched doodling doodlebugs whilst they exploded outside. All the other children were already engrossed in shy silence, busy under the aged and benign Miss Beer's delighted grin, fumbling things from modelling clay. Her granny fingers struggled, but tore me a chunk from a huge bolus, handed it to me and told me to sit down and make something, whatever I liked, and what I liked more than anything in my whole wide world was Tyrannosaurus. rex.

My fifteen-centimetre-high dinosaur was shaped after obsessive scrutiny of all the illustrations in my own frayed collection of encyclopaedias, in comics, on tea-cards and the marvellous shelf-fuls of books in my hall of learning – Portswood Library, to which I was happily led every Saturday afternoon without fail. Here I would sneak out of the 'childrens' and into the 'reference' section, lower the heavy tomes silently onto the smooth honey-coloured tables and head straight for 'T', 'T-y', 'T-y-r', and if I got no joy, then back to 'D', 'Di' and 'Dino' where my beloved monster vied for page space with Brontosaurus, Stegosaurus and Triceratops.

There was a satisfying degree of consistency in T. rex's variously portrayed basic anatomies, its poise and pose, but precious little in its detail and this really annoyed me. Sometimes it was green, sometimes brown, grey … it had round, slit or frowning eyes, sometimes it had a crest running down its spine, in other representations it was smooth. How was I supposed to accurately sculpt the tyrant-lizard king out of plasticine if I didn't know precisely what it looked like? What on earth were all these fossil experts and artists up to?

After consideration and some fiddly scissor work to add jaws, teeth, claws, eyes, and to very precisely cut two fingers into the sadly drooping forelimbs of my model, I was moderately satisfied. It wasn't my best effort, not as lizardy as the one sat at home on my windowsill; that was all green, this was a horrid blend of marbled tones, some idiot having mixed all the colours together so the resultant coagulate was mainly orange. Clearly T. rex was not orange.

As instructed I positioned my dinosaur delicately on a table beneath the classroom's tall arched windows amongst a terrible rabble of malformed and grotesque plasticine blobs. Some of the kids seemed to have made amoebae, others melted cars and planes and one fool had even tried to build a spider. 'You can't make arachnids out of plasticine,' I thought, 'they have stiff legs, that's a job for Meccano.' I grudgingly realised that Karen Harris's mump-ridden mamba actually wasn't so bad after all.

When I returned the next morning all of the things we had made had gone. A large ball of plasticine balanced on the front of Miss Beer's desk. Miss Beer had murdered my T. rex. School didn't get off to a good start.

The day we were all allowed to bring our pets into the classroom was going to be special. It was a nice sunny morning and Batty my black mouse had been spruced up for the occasion. He was in his new second-hand plastic cage, it was mustard coloured, had the mandatory wheel and sleeping chamber but had previously been a torture chamber for my cousin's late hamster. Despite my best efforts to revitalise it the wire remained rusty in places but at least it was more secure than the wooden enclosure my father had made … and Batty had instantly, and repeatedly, chewed his way out of.

Sadly the species list for the class was a meagre four: rabbit, hamster, guinea pig and ... one domesticated house mouse, Batty. They all ignored him, they cooed over the 'bunnies' and those chubby fat-faced tailless things whose eyes bulged when you squeezed them a bit, and queued to offer carrot and cabbage to those cow-licked multicoloured freaks with scratchy claws, but not one of the kids wanted to see, let alone hold, my mouse.

By mid-afternoon the teacher finally caught sight of the lonely boy whispering into his mouse cage in the corner and gingerly agreed to let the rodent walk onto her hand in front of the class. Batty promptly pissed and then pooed three perfect wet little pellets, the classroom erupted with a huge collective 'urrgh' and then a frenzy of giggling, she practically threw him back in his cage and then made a big deal about washing her hands. With soap. Then we were all meant to wash our hands, with soap, but I didn't and no one noticed.

With the mouse cage on my lap and Batty quivering in his favourite toilet roll tube I sat idly waiting for my tardy mother, wishing I had a friendly polar bear who would gobble up all their useless pets. I carried him up the hill, all the while tightly hugging the cage lest he attempt another escape, and as a treat was allowed to have him in my room until bedtime. Then it was judged that my obsessive desire to sit and stare at him might prevent me from sleeping so he was despatched to the dank seclusion of the downstairs toilet.

In spite, I lay listening to the incessant trundling squeaks that he wrung from his furious nocturnal marathons until my parents stopped creaking and were asleep. Then I crept down and using my father's sacred torch shone a milky beam through the door to spy on my remarkably athletic companion. He was relentless, he'd pause to sniff the air, whisking with a web of glassy hairs rooted behind his neat pink nose, and then run, run, run, his tiny feet too quick to see, his tail curved up behind him, all to generate a monotonous symphony of metallic squeals – no doubt also a contributory factor in his nightly solitary confinement.

Batty was the most important thing in my life, but in truth, I didn't really want a mouse; as the name suggests what I really wanted was a bat. I had spent hours pacing the garden staring skyward, hoping to glimpse one, had snuck out of the gate and crept down the road to be closer to a massive tree where owls sometimes hooted but these enigmatic creatures only ever fluttered over the pages of my *Ladybird Book of British Wild Animals*. On page eight it asserted that noctules, or great bats, 'come out to hunt just before sunset' and 'you may be lucky enough to see them in spring and summer' – sadly not in Midanbury.

Eventually, after exploring a number of what I decreed were definitely 'batish' locations – churches, an old school and our loft, which we clambered into several times a week – my father took me camping in the New Forest. We pitched the tent in woodland beside a stream and as it got dark peered over the bank. And then, as he stroked the searchlight slowly back and forth, we spotted some real live bats! It was amazing. I was allowed a go with the torch too, and one of the tiny superfast things flickered just in front of my face. I nearly burst, it was the best thing ever and it made me want one even more.

I had owned a number of floppy 'Made in Hong Kong' bats, which wobbled on fragile cords of shredded elastic and whose rubbery smell was more enjoyable than my vain attempts to get them to look in any way realistic. These crude black blobs were a regular purchase from Portswood petshop, my favourite place on earth and site of a weekly pilgrimage when library duties were done.

It was a short walk down the busy Broadway, if I wasn't dragged into boring Woolworths or dingy Hills the toy shop, which was jammed with too many prams and bikes, past the very dull jeweller's and the wedding dress boutique where my mother would inevitably pause to gaze wistfully at the display and across from Andor Arts where on rare occasions I would accompany her to buy a posh ornament for some luckless relative's birthday. This emporium was carelessly cluttered with a polished fauna of precisely nothing interesting, ever, despite the fact that it was all obviously 'very dear'. Thus I was strictly and repeatedly reminded that I was not allowed to touch anything, because if a vase, bowl, figurine, candelabra or any other gilded trophy were to tumble I'd have to pay for it forever out of my 'pocket money'. Which I didn't even get. In all the years of trailing my quietly 'oohing' and 'ahhing' window-shopping mother round this minefield of gaudy overpriced bric-a-brac I never once saw anything there I wanted to own myself. But across the road, past Alec Bennett Motors, I could have spent everything I'd ever earn, ever.

Portswood Pets and Aquaria had the usual fare: puppies, kittens, the rabbits and rodents, goldfish, budgies, hundreds of terrapins and tortoises, but they had real animals too. Pirate the blue and gold macaw was 'not for sale' but sulphur crested cockatoos were, mynah birds were, and so were flights full of amazing exotics which I would try to identify by matching their biro-scrawled name to the skittish explosions of feathers that would nervously erupt when I peeped into their cramped cages. Their wings whirred like shuffled cards, their beady eyes flashed brilliant fear, the perches crowded with fluxing flocks of Gouldian finches, Java sparrows and pin-tailed whydahs left me awestruck.

But even better than the birds were the tropical fish at the back. These 'ridiculously expensive' things, as my dad called them, made my colouring sets look wholly inadequate; in their verdant pools of waving weed they sparked parts of the spectra I'd never seen, they zipped out of the shadows and flashed brand new colours. They were entrancing and standing in the muggy, damp, bubbling room was tantalising, it was a portal into another world where all the life swam and couldn't be touched, juggled or jostled by my all too eager hands.

Hung like earrings on ribbons of swaying green, coiled and curled, there were once living seahorses and I nearly burst! These peculiar fish ranked alongside bats, otters and snakes, they were ultra special. But they only had them on one occasion and I wondered if they had died because at the front of the shop, by the till, they sold dried starfish and seahorses. And these fragile and easily lost curios were my must-haves or there'd be serious repercussions at home. Just like the plastic bats, desiccated marine life was a necessary zoological sedative to keep those long hours between Saturdays bearable for my parents, so my father always gave me half a crown for a new seahorse.

But even the kaleidoscopic splendour of the tropical fish couldn't match the magnetic draw that pulled me to my knees to gaze into the rows of grubby glass tanks that contained … reptiles! Fat and wrinkled pythons, the peeling tails of boa constrictors and, amongst a fluctuating show of lizards, geckos and skinks, fabulous green iguanas – and unbelievably they had real chameleons too! I'd stand entranced as these jungle gems wobbled through their foliage, eyes going everywhere, just dying to see them change colour or spit out their famously long tongues. They did neither but it was here that my craving became insatiable because the penultimately most desirable pet in the shop was the crocodile.

Yes, in a long thin tank near the counter they had a caiman just like my stuffed one. It lay there, quite motionless in a thick green soup, staring with its glistening, exquisitely veined eyes, its little white teeth so close on the other side of the misty glass. It was beautiful, a pocket dinosaur beyond my non-existent pocket money and my parents' purse. I begged and begged for that creature but not nearly as much as I pleaded to own the greatest animal they ever had for sale in this repository of natural wonders.

Over the years Portswood Pets had bushbabies, chinchillas, chipmunks and even small monkeys – and allegedly once upon a time, lions. None of these came close, because on one fateful and unforgettable rainy afternoon, with my sister in the pushchair and my mum moaning about having to wait outside, I stepped in and there, in a parrot cage, hanging, twisting, twitching, licking with sherbet-pink tongues were two fruit bats. Ahhh! How my father's heart must have sunk when he saw them; I'm surprised he didn't break down and weep because in that instant he would have known that for the foreseeable future his life would become about as unbearable as it's feasibly possible for a mono-minded compulsive child to make it. His son didn't want the grand wooden boxed set of Meccano, the Action Man deep-sea diving outfit, the complete array of Thunderbirds toys, Scalextric or a brand new bike … he wanted a bat. And he wanted that bat very, very badly.

The Bird

Sunday 1 June 1975

UNFALLING, THE BIRD stands chopping air, fluttering and then rolling down smooth, slipping and then sliding away to ring a curve across the storm until it pitches at its apex and begins to dance with the wind, its plumes constantly shaken, folding and flicking to steer it still and . . . balance broken it tumbles and steadies with a twist of grey – cloud-licked and clean, now measuring the weight of the sky again. Then a drop, deckling wings furling – waiting, rich brown back and freckled front – watching, and then the ground quickly surges up and swallows it into the scrolling grass, sucks it down in a greedy rush. And it's stopped, nothing happens now.

Wind licks little furies on the meadow and tousles the willows' petticoats, which flounce into a fit of wild fretting. The evening is set to argue, it's begging for thunder as it cleaves the sun through the groaning barricade of carcassed elms and the drenched field in which the boy lies sours around his torn plastic shoes. But he is unshivering, he needs to see what happens next or all his sodden trudging will have been a waste of Wednesday . . . the bird still in hiding. For him there is no time here, nothing measured, nothing that passes, for him nothing is felt except his indestructible focus. His world is small and shrinking fast and this leaves him alone in the fields of fourteen, tiny in a giant space, safer in himself than in anything of theirs.

And it's up, just there, low, burdened and loping away through the cloak of drizzle. He jumps and runs, runs stumbling and smashes through the spiny hedge, always with his eyes on the bird, it goes up, it's black now, not so pretty, then it twists to a sliver and folds, flaps past the trees where the cows are puddling mud and then higher, it circles tight and once and through the hassle of his panting and the wash of rain he hears a faint whinnying before it vanishes.

It's prematurely dark by the time he slips the elastic bands over his feet and folds his trousers neatly back, jerks his bike around, his jacket loaded with wet and the stale smell of him and the soft scent of earth. He pauses to pull some thorns and licks watery blood from a scratch that is still bleeding when he kicks open the garage door, crashes his racer into the blackness and smells his dead dinner beneath the grill. Friday, he thinks, Friday… his hand on the door, his mum shouting up the stairs, his sister ignoring her, his dad reading about the referendum, ignoring both of them.

The Farmer

June 1975

THE FARMER AT West Horton lopes out watching the mugs slopping tea, scalding his muddy palm, dodging between the dazzling sunlit sheets that his mother stands pegging to a droop of lines. Unhitching a rose-snagged corner, he dirties it and frowns. The tidy old woman takes the tea and crouches to place it by the basket; she smiles, draws curtains of grey hair behind her ears and continues to hang up the washing. He glances out over the valley and sees a tiny figure moving across the jigsaw of meadows by the railway.

From the edge of the garden he peers, slurps his bitter brew and nods. It's the boy who comes looking for birds. He'd been at the door before eight this morning but he'd ignored it; he'd given him permission to go down there but whenever he came, which was several times a week now, he always knocked on the big front door anyway. He'd answer it sometimes, nod and say, 'Fine. That's fine, yes, go, you're welcome.'

Occasionally after this reiterated exchange the boy would suddenly start to tell him about some bird or other. He'd talk absurdly fast, obliviously tripping through his words, always looking down at the step, he'd tell the mat about something that totally switched him on, he'd lurch from timid and backward to a barely contained mania, rambling too quickly, excitedly crashing through a dialogue that gave no room for conversation and then, inevitably, punctuate this cascade of unsolicited enthusiasm with a question. He'd finally glance at him to ask if he'd seen a 'whatever-it-was'. Which he hadn't because he knew nothing at all about birds.

A clod of shit fell from his crusted boot, so he dragged his instep over the rocky edge of the rose garden, drained his cup, flicked away the sludge and then crouched to push the dung around the wizened stems of the flowers. His mother's shadow eclipsed the soil and he knew exactly what was coming.

'Lavington,' she wheezed, 'I won't be able to do this forever.'

The washing. She meant the washing. The housework, the cleaning, the wife's work, the non-existent wife's work. The farmer's wife who wasn't there. He stood up. The kid had stopped by the three big trees on the edge of his herd of curious Friesian heifers, which were crowded round the gate in the bluish wash of mid-morning shade. He stropped his hands over his checked shirt, folded back his tattered cuffs and clipped his thumbs onto his hips. A radio dribbled the Osmonds over the lawn and the distant cattle bucked and jostled, encircling the figure until he disappeared. He wasn't going all the way down there. He screwed up his eyes; it was a long way off. Thankfully the boy reappeared, silently clapping his hands at the playful crowd, and then flipped over the gate and out of sight behind a plait of hedgerows, still foaming with suds of browning blossom. He shook his head, turned and limped off to the yard. Kestrel, that was it, the bird he was always on about. Kestrels.

The Bird

Friday 6 June 1975

'I FOUND IT!'

Blue biro. I'd tried so hard to inscribe it in my very best curly handwriting but the quality of my calligraphy decayed rapidly and after seven short careful lines impatient delirium had annotated the remainder of the entry and I'd produced a page of inappropriate scrawl. I sat up and breathed and, mildly calmed, gave the second whole page over to a neatly drawn map of a tiny patch of Hampshire. It showed individual trees and there was a scale and a legend that illustrated the symbols for fencing, a bog and a wet ditch, and at its centre a great ballooning oak labelled 'nest-tree'.

But this was all a waste of time, a formulaic exercise, because I knew I'd never ever need to be prompted by this diary at all, I would remember the moment I found that nest in unfailing detail forever. That surge of raw ecstasy as the male Kestrel flew into view and his mate squeezed out of her cubbyhole had made me physically shake. And that burst of joy had only slowly transformed into an enormously triumphant and vain crown of satisfaction swollen by the smug knowledge that all my hours of cold wet searching that spring had paid off.

But then I had been optimistic that Friday 6 June 1975 would be 'the day' based on the observations made last week when I'd tracked the hunting male close to that particular spot. Nevertheless, I'd still been curled amongst a prickly posy of thistles, swooning in the bouquet of trampled hay and warm sweet dung for three hours before he'd returned, and there hadn't been a peep or a glimpse of the female. She too had been lying low, brooding her six grey downy young in an old crows' nest, shielded by a thick veil of ivy about thirty feet up in the oak. But once I'd found it all I really needed to know was could I get up to it?

I'd had a powerful urge to sprint over and start to clamber up but from somewhere I'd summoned some control and dutifully remained hidden in the shadow of the hawthorn, fidgeting on a spotty bed of its mouldering confetti. As the morning warmed, I'd gazed unblinking at the bushes opposite, bunched across my horizon like a row of freshly permed heads in a cinema, waiting, aching for the next round of Kestrel action.

Three noisy exchanges had marked the arrival of food, the male performing a delightful quivering flight and soliciting the female from the nest with a shrill call before she disappeared to feed the invisible but now audible chicks. Afterwards she'd swooped down onto a fence post to preen and taken a bath in a shallow ditch that ran beneath her favoured perch. I'd perversely fantasised that this final leg of my quest would take all day but it was only eleven o'clock when it had all happened so I'd pried open my Tupperware box and picked apart a Wall's steak and kidney pie, nibbled a hard green apple and sucked a Mars bar as slowly as possible. This stalling couldn't work for long so I'd packed the rubbish, my binoculars and field guide into my A.R.P bag and dodged the dung-mines as I hurried across the pasture to the base of the tree.

I had made that climb in my daydreams a hundred times, feared the difficulty of the ascent, the sheer height and precarious location of a nest set in the flimsy sky-scraping branches of a giant tree. It would be an eyrie, maybe ropes would be required and obviously I'd be alone as secrecy would be imperative, no one else on earth could ever know the location of such a treasure. I had envisioned it as a rite of passage but although I had written that it was 'quite a hard climb on a thin branch', it wasn't, I'd exaggerated. I had to shin up a bough wrapped with thick ivy with no handholds for about twelve feet but that was easy.

My knees clamped painfully around the branch, my filthy grazed fingers wiped thick brown dust from my sweaty face and then reached forward to part the ivy. I'd hesitated, I could smell them before I could see them, a dry, slightly meaty, warm scent, then I hoicked myself up and leaned forward spitting out a twig. Bright-eyed, blue-eyed, with smoky coats of fluff, they flailed featherless wings and hobbled on fresh custard-yellow legs, their taloned toes tightly fisted. Six, precariously ringing the far side of their stinky platform with gaping mouths, fixing me with terrified stares. They rocked and shuffled, I rested still, not breathing, dizzy after my rapid scramble to their scruffy fortress. I gasped – they flinched, they scowled – I smiled, and gently backed down. They were the most beautiful things I'd ever seen and one of them would be mine. I was possessed.

September 2003

He'd not bothered to disguise his fear. He had pinned himself into the back of the chair as if it was teetering high above the ground, his blue eyes fixed on something atrocious, staring downwards in shocked confusion.

'What did you do?'

After a very long pause he sighed. Then following a deep breath and another sigh he stated quietly, 'I counted them.'

She was conscious not to move, she kept her hands together, her arms folded. She was physically comfortable. Relaxed, composed. But she was pin-sharp now, concentrating on concentrating. Thinking about timing, breathing. She deliberately lowered her eyes to the floor and uncrossed and re-crossed her ankles.

A minute passed. And another.

'How many were there?' she asked gently, but matter-of-factly.

He replied instantly as he always did if the question was objective.

'Thirty-nine.'

She pursed her lips, nodded, and pondered his rationale. He was not rash, he was obsessively self-controlling, but he had also presented a paradoxical collusion of the considered and the recklessly unpredictable. But she already suspected his unpredictable had always been carefully deliberated. So, he would have likely Googled the drug's name and researched its efficacy, he would have calculated the collective potential at his disposal and undoubtedly the only reason he was sat here now was because there had not been enough tablets in the jar.

2

The Tadpole Spoon

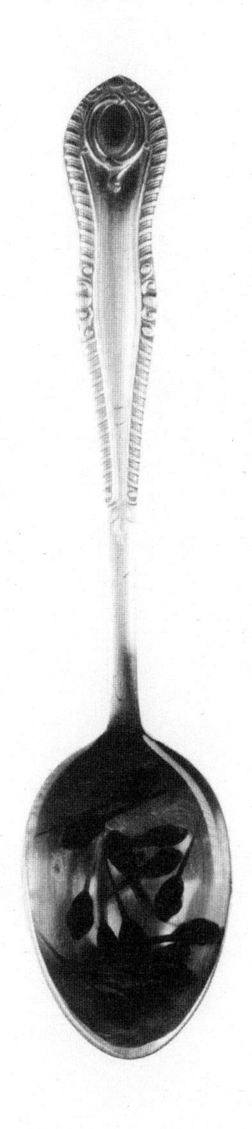

The Old Soldier

May 1973

HE RESTED ON his fork and fumbled in his mossy overcoat pocket for the last few twists of tobacco, which he shook into the corner of a rust-freckled tin. He exhaled and peeled a tissue from under his unkempt moustache and shook the crumbs to make a dusty cigarette, which leaked a ghost of smoke to briefly haunt the shade below his large floppy cap. A lattice of cracks spidered the fingers that gently smoothed the flaked paint of the tin's lid and lifted it to his nose, where he smelled his smoked history, and to his ringing ear, where beneath the whine he heard the distant popping of shells behind the dunes of El Alamein. Thick black genies rolled from the fiery cores of screaming tanks, up-dancing big and fast feeding a sickly stain to hide the sun. And in the torn light of morning he'd cracked the sand-glass that ringed their gutted carcasses as he tiptoed through the charnel, a handkerchief pressed to his burst face, his nose pinched against the warm taste of fresh crackling and it was there he'd stooped for the shiny tin, unburned and brilliant.

At his sleepless bedside on the ward its lid twinkled for two years whilst he listened to the voice of the gun, mocking his deafness with whoops and wails and cracks and thumps, replayed without respite. He lay there, the light pouring out of his black eyes into a world that could never see into his.

He had the appearance and demeanour of a St Bernard: wet pink-rimmed eyes, pinned lobes of skin drooping loosely about thick folds of jowl, coarsely shaven and flyspecked with moles, and with wispy hair that bushed in wiry sheaves down to his shoulders. 'Tramp', all the children called him, he lived in the 'tramp house' and they saved their litter to throw amongst his carefully tended fleshy flowers, thickly planted in polygons of soil set between a geometry of narrow concrete paths. They kicked his green picket gate and tried to pitch their paper and plastic packets onto the roof of his olive-coloured Morris Oxford, which stood permanently clean on two corrugated ramps that ran down a steep sideway to his garage, the doors of which he couldn't open unless he first moved the car. Which he did, each morning at half past nine, lavender puffs jumping from its exhaust as he tickled it to life and reversed it out to park it meticulously along his kerb.

When he drove it was comically slowly and he invariably left either his hat, gloves or tobacco tin on the roof above the driving seat and the skirt of his long coat trapped in the door, a triangle of tweed dragging on the tarmac as he chugged along Cornwall Road and up the slope of Portview Way. Mothers who had walked their children to school smirked at 'the old fool' and youngsters pointed from squeaky second-hand pushchairs as their parents turned to each other to whisper 'shell-shocked'.

He'd hidden in a job, had mumbled down-looking between the shelves and shuffled around his hermitage with a chair and china and a ticking clock that counted his tea times and told him that it would never end. At the bottom of his garden a giant oak cast a shadow map, which mellowed as the evening came to his patch of earth – rhubarb, the glowing leaves of a row of lettuces and the vinous tufts of beetroot all tended and neat with seed packets pegged to the rows, and a water butt humming with gnats. He stood inspecting the wasteland beyond his fence, a great heave of bramble, the oranging banks of the sandpit, the woods and the pinky brick flats into which the shouts of children were sucked as the streetlights flickered and blackbirds rattled their alarms into the scent of honey.

He liked the dusk. When the spent day gave up its light and life to the darkening sky and all the smells came out he'd stand stock-still on his path drawing the cool into his scorched chest, watching the clouds dissolve, smudges of white that gleamed gold and faded imperceptibly to charcoal as they died around the rosy moon waiting for the prettiest stars to prick the blue above the fireglow of the city.

On the hillock above the sandy hollow he could see a familiar form crouched in front of a down-flecked willow, a square sooty patch with a pink blotch of face, crouched and still, facing down the overgrown path into a place hidden by the briars where a spring muddied the ground and leaked into the copse beyond. He had missed the boy arrive, passing silently, only his head visible above the leaning fence, and then stalking cautiously up the slope, creeping and eventually crawling to his dingy spot. He must have been there for twenty minutes or so, he came when the evening flared and fell and now in the squinty dim was only visible because the old man knew where he would be; if he turned away he'd vanish into the twilight.

And then he could just hear them, the faint yips and whinnies, quick in the stillness, piercing as sharply as the stars above the revving engines and crunching gears of the last buses grinding up Woodmill Lane and gasping in relief at their stop on the hill. He tilted his ear to the chips and chirrups and by the time he turned back the figure of the boy had blurred away, he was gone, like everyone had gone.

The Foxes

May 1973

ESCAPING OUT OF my bedroom window was quite tricky. I had to trampoline over the pillow decoy of myself moulded under the blankets without distorting its slumbering form, sit on the narrow sill whilst I closed the frame behind me, then stretch my feet over to the edge of the roof above the saucepan cupboard, and with one shove bridge this divide, slide down its clattering tiles, roll over onto my belly and hang on the flexing gutter to drop with a satisfying slap onto the cement outside the back door. The physical sizes of and distances between the various elements meant that such a feat could not be successfully negotiated by anyone too small or too big, but at the age of nine this mission was not impossible.

There was also an accompanying behavioural strategy to be implemented well in advance of the acrobatics. At teatime, six o'clock, I would announce an illness in embryonic form, a mild nausea, a slight lethargy, symptoms that I would gradually exaggerate until at around eight when at its peak I would be so sick or exhausted that I would have to go to bed early. I cunningly imagined that such an announcement would be a welcome revelation from a hyperactive child, normally still irascibly avoiding his bedroom until ten and any sleep until midnight.

I pictured my parents heaving an enormous sigh of relief and seizing the opportunity to bicker even more venomously. But years later they told me that they had seen through my antics from the outset and took turns to wait up for my noisy scramble back up the tiles. Apparently the note on the door that read 'Please do not come in. I am very tyred/sick and do not want to miss school Tuesday (crossed out), Wednesday (crossed out), Thursday (crossed out), Friday. Don't wurrey I'm okay. Goodnight (drawing of a fox)' was a dead giveaway.

From my hidey-hole on the hill that sloped up to the council estate I was shielded from the path, which fed two sandy lines down to a gully where they divided a huge hedgehog of brambles from some alder trees growing in the wet culvert at the edge of the woods. Beyond was the field, a golden mane of hay dotted with bushy oaks, and beyond that a galaxy of bedroom and bathroom lights, tiny yellow rectangles and squares sprinkled over the ashtray of the city, orange chains of streetlights, car headlights, buses disappearing behind dark blocks of empty offices crowded around empty car parks, horns and the squealing brats and screaming women in the smelly flats.

After the sun withered and the sky got tired the dog walkers hurried home and only an occasional straggler would slip through the blanched chill over the grassy quilt and down under a giant oak and up to the houses behind me. Alongside that huge tree he would be haunting his garden, the smoking scarecrow, staring out into the scrub as the detail decayed, as the blackbirds strafed him with their fierce clinking. The other kids called him 'creepy' and a 'tramp' but whilst they taunted the old man I wondered what he did, what his days were made of behind the dirty windows of his lifeless house. And now couched in my nest I studied him and even from a distance I could sense something massively wrong, something lost or damaged beyond repair. He existed in a world I couldn't see and as he blurred until only the crimson star of his cigarette held him in the murk I was frightened by how anyone could be that far away.

Sometimes I'd imagine I'd heard them yapping in their spiky cave, or I'd think I'd seen a snout poking through the nettle curtain or fancy that one of them had bolted across the tramlines of sand, all fictions of my anticipation. If I was lucky and no one wandered down the hill then they'd arrive earlier, in better light, tumbling into their play as rusty dusty cubs instead of ashy silhouettes that soon turned black and vanished. And they didn't seem to sneak out as much as burst from the bank into a whinnying bustle of rolling legs and tails, gaping mouths, pouncing, bouncing, then freezing and fleeing at the bark of a dog or the owner bawling its name. Then a nose, two, four ears and then they'd all five fall out onto the path and stand gazing towards the noise for two shakes before the irresistible urge to nip each other saw them merge into a furry ball, a free-for-all cartoon fracas.

The vixen would break cover more cautiously, at first just her head, locked looking up the slope, sometimes right at me and I'd hold my breath, unblinking, until she tiptoed out to stretch and sit, studiously ignoring the goading of her litter who would dash behind and beneath her, pushing and shoving. She'd slant her head to scratch her chin without standing and lift her paws to lick them, or twist round to nibble her rump and then stand and shake, before sitting again and twisting her ears, one forward, one back, and then yawn, her tongue curling, her lips peeled to bare her teeth, all the while the cubs jumping to nuzzle her mouth or rooting furiously beneath her to suckle, nudging her to seize a nipple, tugging until she stood and sauntered confidently into the woods. They'd start to follow but without any conviction so would soon reappear on the path, biting each other's rumps and legs, gaping and gently gnawing each other's faces, rolling and running to escape or attack, their tails waggling as they faded into the thicket and tumbled into the snog of darkness.

After a week or so I began edging closer to the fusty hole on the bank where they emerged. Each night I found a new position to hide, where propped on my elbows I could peer above the fringe of weeds to watch, until finally I summoned the courage to wrap myself around the thin trunk of an oak bush almost opposite the den.

I gingerly snapped off a few leaves that might restrict my view and waited. But I'd arrived far too early and was soon tortured by pins and needles and aching arms. Of course fidgeting was impossible; I had to remain immobile and barely breathing, I had to remain dead still, coiled like a lizard amongst my camouflage of leaves. A beetle trundled over my sleeve, ants zigzagged tickles over my legs and a grasshopper scraped a slack song beside my cheek until finally the thrushes heralded dusk, my anticipation abolished pain, the whole world shrank into this one small stage and all that existed was my immense excitement and arthritic focus.

And nothing happened. The nook in the grasses swallowed the light, there were no sounds, I began to fear that they had gone, had died, had been killed. But as the minutes slugged by I knew I had to wait whatever, because if I left, even moved, it could ruin everything, they would see me, smell me, hear me and all would be lost.

And then it was just there – a cub, standing in profile looking along the path. It made no noise; it dipped its nose to the dirt and then slowly sat down, twisting, tensing and relaxing its big ears. It was at most two yards away, too close, terrifyingly close. I froze, I could feel my tongue, my mouth filling with saliva, my teeth clenched, I felt myself blinking, I could hear myself blinking but not a cell in the rest of my body flinched. Another cub materialised, ambled past the first and away but I resisted the urge to turn to follow it. Then I could hear them in their brambly maze and then the fawny ruffian swivelled and stared straight at me.

It had definitely seen me; its triangular face focused on mine and then it very tentatively rose and began stalking towards me, softly, so softly, each tiny paw paused before treading, its muzzle lowered, its black eyes fixed. It came to the edge of the track, its head level with mine, until it stopped an arm's length away. Button nose, sleek cheeks with a fuzz of fine black whiskers, a fluffy, smoky coat, and after the most amazing second of my life it turned and skipped through the hole, its glowing tail tip following it forever in my memory.

It was indisputably and absolutely the most beautiful thing I'd ever seen. And a secret too – my fox, and I'd think of it all day, tapping my biro through the drudgery of double maths, blanking out French to constantly replay that cherished encounter, checking the sky and hoping it wouldn't be raining at fox time. But the times and the cubs soon became less predictable; within a fortnight they had dispersed and would only appear alone or in pairs for a few moments. I tried luring them into view with food – the carcass of a whole rabbit my parents had bought me from the butchers, which I had skinned and skulled, and road kill that I collected on my bike rides that I wrapped in bags and secreted at the bottom of the new freezer. I'd wait to see them drawn to my bait but only find them at the top of the field as I snuck home and on the last twilight I stole from that summer I saw none until I got back to the gate and, as I clamped the catch in my palm so it wouldn't rattle, I watched a cub flow along the bottom of the road roving from silhouette to shadowy amber through the streetlights until it stopped by the old soldier's house, glanced up at me and then slipped down his driveway into the black beside his car, which I noticed still wore his cap above the driving seat.

The Bird

Saturday 14 June 1975

GOLDEN WONDER. THE scarlet script was barely visible, the interlocking letters forming a familiar shape rather than discernible words whilst 'Cheese and Onion Flavour Crisps' was just a blur above the carpet tiles. I listened to my breath, to my heartbeat through the mattress, to the house waking as the gloaming gave some form but little colour to my crowded room. With one eye open I lay examining the space – the cardboard box and a pile of books, amongst them the spine of my falconry manual showing the back of the author's head, Humphrey ap Evans and the publisher's name, John Gifford.

I couldn't make out the actual print in the murk, but I didn't need to; like every word in the book I knew it by heart. Its dedication to Saint Bavo of Valkenswaard, the patron saint of falconers, the foreword by His Grace the Duke of St Albans, Hereditary Grand Falconer of England, the picture of a youthful H.R.H. the Duke of Edinburgh with a Peales falcon, in black and white, facing right, left hand in pocket, the falcon peering at that hand, probably thinking it held food, the Duke looking confused; it was all there in my mind, every last page, paragraph, sentence, all the diagrams, I liked those diagrams and the photographs. I loved the bit inside the jacket that read, 'The author shows clearly that it is not essential to possess unlimited means and leisure time for the enjoyment of some form of falconry', but most of all I adored the title, *Falconry for You*. Because today falconry would be for me. That crisp box was lined with hessian, as directed, page 81, and at some point this afternoon I'd be putting a young Kestrel into it, and that's why I hadn't slept all night.

I got up, crept around the touchline of my Subbuteo pitch, crippled Kevin Keegan and quietly switched the light on. The bottom drawer of my egg cabinet was lined with green card and the equipment I'd acquired was carefully laid out as if prepared for one of the book's illustrations. Four jesses and four bewits neatly trimmed from a panel of kangaroo leather that I'd received from my uncle in Australia. On page 49 of my guide it said that dog leather was the best, but that baby seal, porpoise, white whale and kangaroo had also been tried. I'd plumped for marsupial skin and cut the little straps from templates I'd made from paper and I'd then practised tying them on the shaft of a pencil as a substitute for a falcon's leg.

Next were the three swivels and the six bells I'd imported from Pakistan, arranged like trinkets in a jeweller's window, precisely positioned with Blu-tack. I teased them free at least once every day, they were wonderful objects, precious, and had been so hard to come by.

I'd visited the Falconry Centre at Newent and tried to speak to Phillip Glasier, the owner and a renowned falconer. I wanted to know where he got the bells for all his hawks and falcons but he was apparently 'too busy' to speak to me so I'd later written to him instead. He wrote back, two lines, saying he didn't have any sources for 'hardware' at this time. I didn't understand, he had loads of birds all tinkling away, I'd seen them. I wrote again and he replied, one line ... no bells. Having convinced my father that a second trip to Gloucestershire was 'essential' I wrote a third time explaining when I was coming and asking if I could meet him briefly for some advice about training my first Kestrel. He didn't reply this time but when we turned up and my parents were having their packed lunch I snuck off and went to his house.

A stern-faced young woman answered my tentative tapping. I was exceedingly nervous but managed to splutter out my story and explain my desire to please speak to Mr Glasier, if that was okay please, thank you. She disappeared for a minute and then cracked open the door and told me to wait. So I stood patiently listening to voices inside, laughter, then crouched as a dog scratched at the inside of the door, and then finally sat watching a tiny red-thighed falconet bounding back and forth across a rusty budgie cage in the shade of the messy porch. It was scared, scruffy and small, it couldn't settle and when I knelt down to study it I saw confusion and desperation, a sadness in its peeled forgotten eyes.

The house was silent and the mini-raptor still frantic when my father beckoned me from the gate. He was furious, they'd been 'looking everywhere for me', 'been round the whole park twice', 'your mother has been bothering the staff', he spat, it was 'time to leave – now', 'the traffic would be bloody awful'. Apparently I'd been gone two hours.

I picked up one of the silvery bells and cupped it in my palm so it only made a dull clicking. A postal order sent to Mohammed Din & Co, Hawk Merchants 21, Prem Gali, 5, Railway Road, Lahore, the address gleaned from page 54, had seen a small cube-shaped box handed to me by the postman. The whole thing had been sewn into a muslin sheath and decorated with loads of exotic stamps and labels, but when I'd sat on the stairs and eagerly shaken it there was only a disappointing rustle from inside. Each little bell had been individually wrapped in strips of newspaper and a sliver of cloth fed through its slit to stop the little lead pellet from jingling.

I just couldn't understand why Phillip Glasier wouldn't tell me where to get the equipment, it didn't make sense; today I was going to become a falconer, like him. It wasn't very nice to ignore people who really wanted something, especially when you had it yourself. But then Mrs M. Paton and E. Gibbs (Mrs) at the Home Office, Romney House, Marsham Street, London, SW1P 3DY didn't seem to want me to be a falconer either.

I'd dutifully followed the advice, page 199, and applied, through my father, for a licence to remove a Kestrel from a nest. The double-sided application form asked in section 7 for 'Particulars of any club or society concerned with falconry which you are a member of and particulars of your experience in falconry.' It added, 'If inexperienced state the name, address and experience of a falconer who will assist you training in any bird.' I wrote to Professor E .W. Davis at the British Falconers Club asking to join. After a second letter came his reply: 'I am afraid we cannot accept you as a member until you do obtain a licence for a Kestrel and are prepared to produce a copy of it'. Surprisingly he knew of no falconers in my area, but I made the application anyway. The answer from the Home Office was a pre-printed sheet and read, 'We regret to have to inform you that your application cannot be included within this year's allocation and is therefore refused.'

I fussed with the box, we'd had no old newspapers so I'd folded my poster of Marc Bolan and scrunched it into the bottom. He glammed the gloom. I closed the lid, switched off the light, got back in bed and woke up as my dad slammed the front door on his way to work.

Twelve hours later my bedroom was a different place. It had a Kestrel in it. Perched on my jittery paw. It was gawky, half-dressed, its jumper ruffed up over its baggy trousers and sockless feet, an in-betweener with a tetchy temper, tufted with sneezy down. I could smell it. Sweet, musty, dry and when it shook, a cloud of glittering dust puffed into the shaft of evening sun that cut through a crack in the pegged-together curtains. I could smell its droppings too, or mutes as falconers call them; wet and papery, they had slapped Marc hard across his starred cheek, blotted his sparkling corkscrew hair and blistered his guitar.

I was sat dead still in the armchair we'd hefted upstairs to my bedroom and when I did need to shift I moved Six Million Dollar Man style, in ... slow ... motion. I was 'waking' it, preventing it from sleeping until it agreed to sit on my fist, which it was doing, until I moved, even slowly. When its head dipped and it began blinking its eyes closed, I gently rocked my hand or tickled its toes with a pheasant's feather and it sat up startled and every now and again threw itself off in a bate, a fit of fruitless flapping. After a few hours I'd begun talking to it in the softest whispers I could draw from my state of paralysed excitement, but I'd run out of things to say so I'd been reading *Brave New World* aloud to get it used to the sound of my voice. As recommended on page 89.

To my immense relief all six eyasses had still been in their eyrie when I climbed up in the morning for a preparatory peep. But when I'd returned mid-afternoon with my dad I'd needed to scramble a yard further up and balance precariously on my knees so I could use both hands to grab the one I wanted from the gang of gaping louts as they fled to the brink of the fly-buzzed and flattened nest. I was worried they'd all tumble out so I quickly bundled the skinny little thing into my A.R.P. bag and then trembled down the tree, strode after my already fleeing father snorting dust and pulling twigs from my hair, before transferring it to the Bolan box and then sitting in giddy silence as we made our getaway in a 1500cc, 74bhp, coffee-coloured Austin Maxi.

Once at home my father had regained his composure and held it down wrapped in an old towel whilst I'd attached the jesses and bells, the pencil pre-training not counting for much. Then I'd fed it some beef, banished my sister from the room and settled down for what my mother assured me would be a 'long old night'. But every minute was magical, every single thing it did was fascinating and everything it didn't do was equally wondrous, and to be sat there, with a Kestrel, a real live Kestrel, my own real live Kestrel on my wrist! I felt like I'd climbed through a hole in heaven's fence, like something shiny had fallen and I had caught it with my heart. I reeled through the night watching two, three, five and seven o'clock wind by on my Westclox and mapped the future of our lives in my head, planned everything and reran and reimagined it until it was all absolutely perfect.

The Old Lady

July 1975

WHEN THE BAKING afternoon sun chattered through the drooping clusters of rowan flowers, scorching the gap between the pear and the cooking apple tree, the wasps came to strip the silvery surface from her trellis. She steadied herself on the washing pole, her wedding ring scraping over the pitted steel, leaving a wiggly line of gold, a miniature map of her fingers' shaky grasp on the rich red rust. Drawing a deep breath she pursed her lips and listened to the faint scratching of the insect as it backed down the strip of wood in deliberate steps, its yellow-banded jacket panting furiously, its head bowed and too busy to notice her leaning in, the sun catching her spectacles, their lenses projecting a fluxing pattern of white light that quivered on the broken fence beneath her, bending bright lines and a hot pinprick, a sprite, dancing and gone as she turned to look through to her neighbour's garden beyond.

It was all still, the single line of paving slabs glowing, some sparrows bathing and chittering in the lea of a high ivy-covered fence that led away to the blue-framed windows and door, the shadowed glass reflecting the maroon of the peony heads that had ruptured in the beds around the house, beefy globes disintegrating above the frilly pink hem of London Pride. She waited, some doves gently moaned and a returning wasp buzzed around her head looking for the source of its paper work. And then she heard it. A pretty tinkling, a short liquid rattle that reeled in a brief burst and then three or four single softer notes, a sound that cut cleanly through the sultry density of their gardens, was so simply refreshing and now almost imagined in the humming glade where she stooped. She twisted the pearly face of her watch until the fine hands caught the light, sighed and shuffled slowly back up her path, her crepe slippers silent on the swept steps. She flicked the kettle switch, she was too early to meet him, silly, so she would have a nice cup of tea and go back just before four thirty.

He was just coming out of the rickety door when she started back to the fence, the chicken-wire panel wobbling shut with its distinctive clatter. Then he was standing, fiddling with something whilst the bells rattled, the flurry of icy silver filling the space, but he lowered slowly out of sight just as she reached the pile of sticky grass cuttings that formed a bank up to the fence that divided their gardens. Leaning forward she could see him sitting in the shade, his left arm resting on his knee and at the end of it, on his closed fist, a bird. He was gently tweaking it with a feather, using the tip to tease its soft creamy chest, to neatly align the row of dripping spots that ran in thick lines down its fluffy flanks.

It was small, bigger than a starling or a thrush but smaller than the fat pigeons that clapped noisily onto her bird table whenever she shook out the biscuit tin. Overall it was a fresh gingery mahogany colour but it was heavily barred across its back and down its tail with darker chestnut brown, maybe black. Its knife-like wings, crossed over its tail, were darker but edged with the faintest line of tan, which ran down their edges and pooled in a crescent at each overlapping tip. It was bobbing its head and occasionally stooping to half-heartedly snatch at the irritating feather, each movement sending a soft tinkle of metallic notes from the shiny bells that hung from its legs on the thinnest strips of leather. And then – a sizzling crescendo as it bent forward and rapidly scratched its chin, nibbled at its thick lemony toes and stamped its shaking leg before settling back to head-bobbing, turning with quick jerks to look down the path.

When she murmured 'hello' the bird instantly transformed from a cuddly ball to a sleek starved thing fraught with fear and staring with black bulging eyes. It ducked, flinching, its wings half open, then held them drooping at its side, leaning forward, lifting its tail before flinging itself off his fist flapping, smacking the air like a rag in a gale, slapping his arm until it hung wings open, tail fanned, panting, furious and scared. He gently slid his hand beneath its back and lifted it back onto his glove, it spun round to face her all bug-eyed and he glanced up frowning. She moved her fingertips from her lips to her chin and whispered, 'Sorry, sorry. I've just been wondering what was making that sound, the little bells, it's lovely.'

He was looking down at the bird, carefully straightening the straps that held it, adjusting the loops of nylon cord that spooled evenly beneath his hand, the startled creature still glaring at her vengefully.

'What is it? It's beautiful.' She had already noted its little hooked bill and its sharp wings, so she knew it was a hawk of some kind. But Christopher was fixated on his pet, which did seem to be relaxing slightly. He had always been slow to venture speaking but she knew that if she persisted he would talk, certainly about animals. She remembered their first little chat through a steamed-up jam jar full of caterpillars and knew that if she got something wrong, called them butterflies when they were moths, or frogs when they were toads, then he couldn't not correct her and this would generally get him started. She'd have to be patient, so she waited, peering through the split fence at part of the small green cross that he had long ago told her marked the grave of his pet mouse. Finally she cleared her throat and asked, 'Is it a hawk?'

She thought she might have seen him almost imperceptibly shake his head.

'Is it a sparrowhawk?' That was the only name she knew, she could hardly ask him if it was a golden eagle, she wasn't a complete idiot, even if she knew nothing about birds.

'Where did you find it ... ?'

The bird had settled, it was now nodding its head up and down again and looked a bit fluffier, but it was still staring at her, bright-eyed and so alive. He gently unwound his crouch to stand up, holding his arm rigidly horizontal throughout the graceful manoeuvre. She took this opportunity to place one foot on the bank to lean further forward through the broken trellis and noticed he was wearing a heavy army-style coat, he must be boiling she thought, and just before she could ask if it had a name he said in a soft, calm voice, 'It's not a hawk, it's a falcon.'

Bingo.

'Oh,' she said, 'I thought falcons were bigger,' and before she could continue he replied, 'There are four types of falcon in Britain: the peregrine is the biggest, the male merlin is the smallest, the hobby is the rarest and this is a Kestrel. It's a male.'

He turned towards her, put his scruffy plastic sport shoe against the fence and then warily drew the bird up level with her face. It immediately tightened into itself, ducking and peeking over its shoulder for somewhere it might bolt to. He raised the feather and shushed it, stroking its breast, tweaking its beak with the tip, trying to distract it from its urge to flee. But the ploy failed and again the bird flew off his fist and flapped, all feathers and rattling bells until it hung exhausted and he lifted it back to his glove.

'It's called a bate when it does that,' he said, 'It's a falconry term. At the moment I'm manning it, that means I'm taming it. It needs to get used to other people, cars, dogs … stuff … everything, so that when it's flying free it won't get scared and fly off.'

'No, well, you wouldn't want to lose it would you.'

'I'm not going to lose it,' he snapped seriously, 'I'm never going to lose it.'

'No, of course not,' she demurred to placate him, 'it's a precious thing isn't it, it's beautiful.'

'These are its jesses and this is its leash.' He stretched out the straps on its legs and the loops of nylon, 'these are made from kangaroo leather which I got from Australia and this is the swivel.'

'Kangaroo!' she exclaimed a bit too loudly, the bird spinning suddenly from its preening and glowering at her through the fence. They both paused and faced it, holding their breath, fearing another bate.

'It says in my book *Falconry for You*, by Humphrey ap Evans, that it's the best. But I haven't got much, I've got to make it last.'

'What about the bells, where did you get them?' she enquired.

'Pakistan,' he stated emphatically, 'I imported them from Pakistan. Wait here ... ' He tugged open the decrepit aviary door and disappeared inside, rummaged about and then backed out, carefully drawing his fist and his bird through the opening, and then handed her a package. It was a square box, big enough to hold a mug, and was made of cardboard. But the outer layer was a thick formerly white gauze that had been neatly stitched along each edge and plastered with several large stamps and a label, handwritten in English, which read 'Mohammed Din and Co, something or other, Lahore, Pakistan'. From within she could just hear the muffled jingling of some more bells. She shook it.

'Spares,' he said.

She passed it back and he told her quite a bit more about the 'art' of falconry, about how his bird, the Kestrel, was the best species for beginners and that goshawks and peregrines were only for experts, which he would be one day. He showed off the bells on its legs, their polished surface already fading to reveal a dull brass beneath, and explained that he had chosen a pair with a slightly different pitch and tone. This, he said, meant that they could be heard up to a mile away depending on the wind and the height of the falcon, which might have 'raked away'.

Then she heard her front door go, her daughter thank goodness, so she said goodbye to him and set off, squinting hard after her time under the trees. As usual he was an encyclopaedia of intense and exhausting detail, which in truth soon got a bit much.

Halfway down the path she turned and raising her hand to shade her eyes she saw he was still there, whispering to his bird. His face was really close to it and it was bending forward and playfully pecking at his nose and, although she could only see his silhouette in profile, she could tell he was smiling. She smiled too and then went indoors.

The Cinema

December 1966

WE WERE LATE, my mother was hurriedly ramming my push-chaired sleeping sister up the high street, the lights pricking the puddles with twinkles, the colours all smashed and splattered, flashing and scattering as the buses grumbled or careless feet skipped over the slabs sprinkling radiance in their rush. The pavement was beautiful, it had Christmas splashed all over it. The rain-fresh cold smacked the smoking faces pushing past and the brilliance of the shop windows, curled with glitter and tinselled trees, with glistening globes and a thousand Santas, red and white with presents tumbling and cartoon reindeer in starry skies, shone out, broadcasting a palette of excitement so rich that I couldn't squeeze it all in at once. My eyes darted from Thunderbirds to Scalextric, to selection boxes and out to the flickering webs of illuminated snowflakes and angels dancing above and all down the road. And then as we dodged through the bus queues and into the fireglow of the cinema foyer I saw it and everything went awww.

It was unbelievable. Beneath the bright orange title a busy scene was filled with dinosaurs doing battle with each other and armies of tiny people under a rain of volcanic lava. Brontosaurs wallowed in lakes beneath mountain peaks, a Triceratops was goring a Ceratosaur and the meanest-looking giant turtle was attacking over the crest of a sand dune. A Pteranodon was hovering bat-like, clasping a woman in its feet, whilst other flying reptiles soared in silhouettes of green and navy blue and best of all, beneath the legs of some cavewoman, a man was fighting a ferocious Allosaurus. *One Million Years B.C.*

The Usherette

December 1966

SHE PUSHED HER way off the number twelve against a stale wardrobe of wet coats shoving to get a seat, shop girls with their shopping, women steering their toddlers and folding pushchairs and young Brylcreemed blokes, loud and leery, a baby squealing and the conductor telling everyone it was full and lazily lying that there was another one coming along right behind. Her left foot landed in a puddle that would stain her brand new tights with diesel spots. She swore, jostled free of the scrum and dashed into Boots' doorway to straighten herself out, pulling her coat on, doing up a button, hitching her bag strap onto her shoulder and checking that her purse was still there. Then she waited for a break in the traffic and then a break in the bustle before tottering out into the downpour to the chorus of 'Jingle Bells'.

She went straight into the ladies to finish her make-up, sprucing up her glue on lashes and delicately drawing her crimson lips, pouting and checking her profile. Pinching the shoulders she shook out her blouse, undid a couple of buttons, tugged at the cups of her bra to resettle her breasts, twisted her skirt round so the zip was at the back and then began to paw through the line of jackets hooked on the wall until she found hers. It stank of smoke and was dotted with cigarette burns. It was too small but when she did the middle button up it squeezed up a nice cleavage. It was red to match her lipstick.

When she paraded out he was there, pacing up and down the gaudy carpeted slope that led unevenly to the double doors that opened into the auditorium. With the big lights on it looked like a sty and he was the pig. The king pig. Mr Sleaze in a beige polyester suit and loafers.

'Ali', Ali', bloody Janet hasn't bloody turned up, get into the box will you, it's an A so there might be some kids in. And for Christ sakes sell them some bloody sweets will you,' his fag wobbling on his lips, eyes squinting, head back, his whole body wincing.

'Yes sir!' she saluted, 'right away sir,' she curtsied.

'Oi, button your lip young missy,' he snapped, jabbing his fat finger, 'and sort out the bloody mess whilst you're in there.'

The ticket booth was cupboard-sized, when you balanced the cushion on the stool and perched on it, your knees banged into the cash tray and your back was up against the rack of chocolates. A jumble of torn boxes piled on the bare wood floor were filled with coloured rolls of tickets and a nest of John Players butts spilled out of a Watney's bowl onto a shelf decorated with a confetti of toffee wrappers where five half-filled mugs ringed with coffee scum teetered precariously. There was nowhere to put anything, it was impossible to tidy.

She groaned, sighed, snatched up a rusty aerosol of window cleaner and blanketed the glass in front of her with an acrid froth. When she'd smeared it cleaner she looked out onto the high street and there standing alone and stationary amongst the gyrating mayhem on the pavement was a boy. He was gawping up at the poster board, totally mesmerised, his mouth open, his eyes open wider in a state of utter wonderment. He took a couple of steps forward into the full glow of the foyer and just stared at the poster. Then a woman, his mother, crashed a pushchair to a stop and shouted something. Nothing happened. She stood there for a few seconds scowling at the billboard before lurching forward, seizing his wrist and towing him out into the throng. She was angry, that was obvious, he was in another world, that was obvious too, as his gaze never left the picture, even as he was dragged away.

'Bleeding hell, what a bleeding pigsty,' she muttered to herself, craning down to try and find anything to clean up with. There was nothing and when she stood up again the kid had miraculously reappeared, closer to the booth this time. He was in a trance of absolute awe, his hands were clasped in front of his buttoned coat and he was wringing them tightly, his eyes racing around the picture, he was mouthing things she couldn't hear. She watched him for a minute and he had no idea she even existed, it was like he was hypnotised and whispering spells to himself. So, slightly worried, she clambered out and cautiously crouched beside him, careful not to put her knee on the greasy tiles and completely ruin her ruined tights. Together they studied the poster.

It was for a monster film. *One Million Years B.C.* She could see instantly why his mum didn't want him looking at it. There was this tart with massive tits in some sort of caveman bikini, with blowy messed-up hair and fluffy booties just standing there with her legs open. It was so typical of all the pictures Mr Pig got in, never family stuff or romances or kiddies' cartoons, he always ran blue movies or anything that was halfway there or just about legal. She never told her mum exactly which cinema she worked in, the old dear would be ashamed of her, they'd be gossiping over the washing lines about it if they knew she had a job in the Regent. And whenever her lot thought about going to the pictures she'd dread the conversation as they'd get round to asking about which film her place was showing and she'd have to make an excuse to leave the room.

She eventually found the tiny black 'A' certificate hidden in the corner, it couldn't be that dirty then she thought. Nevertheless that slapper looked certain to be losing her swimsuit, no doubt one of the monsters would be tearing it off. She looked a bit like her brother Steve's last girlfriend, only with more muscly legs and a better tan. That bleeding cow had fancied herself an' all, and her old man had been all over her, 'Sylvia this, Sylvia that', anything to get a butcher's down her bra.

Her leg was aching, so she turned to the boy who simultaneously almost faced her. She smiled, he said 'Allosaurus', and grinned back at the poster.

'What sweety?' she asked.

'It's an Allosaurus,' he replied, pointing between the cavewoman's legs. She looked, there was some sort of prehistoric monster having a go at a hairy bloke with a spear.

'Oh, yes love ... an Allo-whatsit,' and then the penny dropped, it was the monsters he was fascinated by, poor little thing.

'What's that then?' She pointed at a pointy-headed bat.

'A Pteranodon,' he pronounced, followed by, 'It's wrong. They lived 85 million years ago, not one million years before Jesus.'

'Blimey.'

Then as she stood up his mother swept in like one of them teranno bats and ripped him out of the foyer in a whirling fury, squawking about 'missing the bus' and hauling him backwards into the night. 'Blimey, she's got her bleeding work cut out with him,' she thought.

The Pond

April 1967

BETTER THAN MY birthday, much better than Christmas, Crackerjack or the Supermarine Spitfire Mk. V, or even going to the zoo three days in a row, was 'Tadpole Time'.

In spring we made a special trip to the ornamental lake on Southampton Common, a large park just north of the city centre, where common toads gathered in enormous numbers. We went on a school night, after dinner, in the dark, parked the Ford Anglia and then walked silently with the torch switched off carrying jam jars and a bucket, the orange pin light of my dad's cigarette glowing as we picked our way between the trees towards the pale tray of water where ducks murmured and a few surly fishermen hunched in the cold. I quietly begged my dad to ask them if I could see in their nets, which stretched out across the gravel and into the silt, presumably holding invisible riches. But he wouldn't, he just shushed my whispers and ignored my short sulk as we slipped past them and squelched out on the slushy path that led to the marshy side of the lake, ducking under willows, him out front in charge of the torch and the big net, me tripping over stumps trying to see the ground between the shadows of his legs, listening to him swear as his work trousers got snagged by brambles. And we heard an owl.

In the lea of an island covered in thick pines the shallows churned with rafts of amorous amphibians, splashing from the surface and diving into the amber water where we spied on them with our weak beam, and in my excitement I scooped up a load of silt and weed so heavy it bent the net we'd got from the newsagents at the weekend, green plastic, white wire, bamboo cane.

Eventually we got a few in the bucket and I crouched over it with the lamp, my dad leaning on a stump smoking, the toads paddling in circles, the big fat females carrying one or more males on their backs. Their heads craned down, pushing obliviously round, the single males were blinded in torch-tortured confusion in the orange bucket orgy, their skin soft, slippery and knobbly, their bodies bony beneath my probing, stroking finger. In the big jam jar, crystal lit and cold, held up to my face, I saw their throats undulating, their rough marbled bellies, their chubby fingers pressed against the glass and their enamelled eyes and pinhole nostrils all mixed in a minestrone of colours from khaki to cinnamon to olive, splodged and splattered with chocolaty spots, with no symmetry but lovely all the same.

I wasn't allowed to take any home, the bucket was just for looking and to carry the 'spawn water', the spawn itself wound into the jars, a milky jelly suspending blue-black beads, slopping in the battery-flattening flashes over my damp thighs, my clammy long trousers stuck to the red vinyl seats and in the mix some curly shrimps and a boatman were struggling amid the viscous phlegm. In through the back door, the prizes lined up and dried on the kitchen sink, wet sock prints on the lino, into my pyjamas and one last look before the light went out, I heard the theme to *Softly, Softly* and the windowsill aquaria seeded dreams of thousands of wriggling tadpoles, ink spots in sparkling water with bright green weed.

Tadpoles were brilliant. Better than mice, otters, bats, seahorses and even things like pythons and cobras. They were free, came every year, lived in jam jars, weren't dangerous or poisonous and if they died I just went to get some more. But I didn't want them to die, I wanted them to turn into toads. Or frogs. Of course they did die, by accident, or if my mum put a finger-sized chunk of liver in with them as food or if I left them on the outside of the curtains when it was sunny. Or if my stupid sister kicked the jar over whilst we were pretending to endure a violent storm whilst playing shipwrecked on a make-believe life-raft bed. Even if I rushed to get a spoon I couldn't get their squashy bodies off the bedcovers in time.

And when it came to tadpoles a teaspoon was, after the net, the most important tool so I had a special 'tadpole spoon'. To the uninformed it would have looked like any other battered piece of Sheffield steel, it was after all only one of several old spoons that were part of a mass of cutlery my parents kept amongst a tangle of other scrap in a sticky uncompartmentalised kitchen drawer.

Every breakfast, lunch, tea, dinner or suppertime this crate would be yanked open causing the multitude of metal things to slide violently back and forth, mixing the forks with sugar tongs, knives with nutcrackers and spoons with centuries of battered silver, nickel-plate, stainless and chromium. Trying to find enough matching items to make up four complete place settings was a tiresome daily puzzle. To achieve the minimum of twelve pieces needed, the luckless table-layer would have to rattle the grinding junk of metal round, peering into its jigsaw to recognise any visible fragments of a telltale handle. I learned to start two patterns going from the outset and mulled over the mathematics as I laid each out neatly on the grey Formica top whilst repeatedly re-enforcing my ambition to have a kitchen, a house, a life with separate sections for everything.

Once a year I'd pull the sagging drawer out onto the floor and kneel over it, flicking through the grimy, gooey items that had gradually gravitated to the back. It was from here that I'd retrieve the sacred 'tadpole spoon'. I'd wash it and rinse it in cold water and place it in plain, simple, clean and neat isolation on my windowsill, lined up nicely with equal space all around it, as befitted any valued tool. It was slightly modified, I'd bent its neck through about forty degrees to facilitate more efficient removal of those tricky tadpoles which clustered like berries just under the neck of the jam jar, a place that normal spoons just couldn't reach.

As a toddler I probably ate tadpoles, especially as I was given bowls to collect them in and spoons to fish for them, but such consumption wasn't deliberate and contributed little to my habit of tadpolephagy. And whilst terminal for the amphibians it obviously did no harm to me. Considered ingestion came later, between five and nine with a peak between six and seven and a half. By this age I was very actively curious, not only about the taste of tadpoles but also about their potential ill effects; having by then experienced stomach ache, flatulence and bouts of diarrhoea, I wondered what would happen if I ate a few. Having returned from school, watched Captain Scarlet improbably survive another absurdly explosive scenario without even the severance of a single strand of the puppeteer's string, jabbed rubberised fish fingers back and forth across a plate until the banana-flavoured Angel Delight arrived with slices of cut banana set in it and a generous blob of strawberry jam dropped vaguely centrally, been allowed to get down and retreated to my bedroom to unnecessarily decant several hundred tadpoles from jar to jar, I found before me, glistening in the early evening light, a spoon full of small, soft, benign, stingless fruits of nature of which there was no shortage. So what could be the harm? And critically when my world was constantly erupting with novel experiences, what, I wondered, did tadpoles taste of?

Muddy water, slightly gritty, strange when one wriggled beneath my tongue and quite tricky to bite or chew. Not like jelly, or sweets of any kind, or like rice pudding or tapioca. Maybe similar to very watery semolina in terms of texture. But beads of softened wheat didn't thrash about in my mouth, unlike ... half-matured toad larvae. And they were quite 'moreish' simply because they were so difficult to taste, easier when I scooped ten to fifteen big ones all into the spoon at once.

The result of my possibly excessive appetite for juvenile amphibians wasn't diagnosed through any medical examination so I can't prove anything scientifically. But these harmless inoculations probably positively contributed to the ignition of a spark that fuelled a lifelong interest in living things, an enduring curiosity in everything that creeps, climbs, bites, stings, slithers, scuttles or slimes; and in entirely romantic terms, I imagined, the molecules of the tadpoles I digested were fused into the fabric of my eyes to facilitate a heightened awareness of life and instilled a profound love for it, the likes of which could never have arisen from my sterile school studies, from the disconnected or imagined experiences that I gleaned from my books and television, but only from that heart that fluttered, as my throat was tickled, softly, by simple beauty at that essential point in my own metamorphosis.

July 2003

'I started to feel indestructible. There were a whole host of reasons I suppose, but I think they certainly changed because I started to feel indestructible. I felt like Captain Scarlet, I had that "Captain Scarlet feeling". You know, you remember him? That Gerry Anderson puppet from 1967? The series that came after Thunderbirds? He was a Spectrum agent protecting the earth from an alien invasion from Mars. The Mysterons were the enemy. I'm sure it was some sort of metaphor for the Cold War, they were the evil Russians trying to invade and pervert the minds of us western "earthlings". I liked Captain Scarlet, I liked the way he dressed, I liked his neat outfit ... But his most admirable characteristic was that he was indestructible. The Mysterons contrived to kill him every week, he'd get blown to pieces, he'd be in a massive car crash or fall off a skyscraper and then there would be a little jingle and he'd be standing there, brand new, undamaged.

'I needed that, I needed to be able to rebuild myself like him. That's what I think it was, and to make it happen when I was fifteen or sixteen I consciously developed this immense resilience. I think that I thought that if I could survive, survive all the mess in my little world, then I could probably survive anything. I needed something to get me through and that's what changed everything, it didn't matter what they thought then, it didn't matter what they did, I convinced myself they couldn't harm me, ultimately that they, anyone, all the Mysterons out there, couldn't

damage me. I used that secret self-confidence, that energy for years.'

'And when did this indestructibility stop?'

He pulled himself up, briefly glanced at her, and said, 'I don't think it has stopped really, I think it's just hiccuped a few times. Like when I was going to take those pills.'

'And what was it like, this indestructibility. How did it manifest itself?'

He looked at her again, half smiled and said firmly, 'Is *like …'*

She nodded an apology.

'Fearlessness I suppose, complete fearlessness, reckless disregard for my own safety, health, well-being. I'd take physical risks with little care for any potential consequences. But far more importantly it was then that I became consciously aware that I didn't care what other people thought, how what I did affected them, or not. I didn't realise it at the time, of course, but this was the point when I must have begun to understand that I wasn't properly programmed to empathise or sympathise with other humans. That's why accepting that I didn't care wasn't exactly a struggle, I suppose. But what was crucially different now was that I'd decided that I mustn't let them upset me back.'

He paused and they listened to the loud whines of a catfight just outside the open window. When it had wound down, he raised his eyebrows, smiled and went on.

'I don't know, maybe I'm wrong about all this, but lots of things changed, maybe it was just me, adolescent me. Maybe everyone was changing like that, but whatever, I clearly went in a different direction, never turned back. It supported my separation and even if it is, or was, all rubbish it allowed me to recognise myself as different at that crucial point, it was the start of me accepting

that I was not exactly like them, that I couldn't be, for all the wishing, fretting and crying. And that was hard. I had no one reassuring me that it was okay. Just them all telling me that I was not invited. So I needed to be indestructible.'

'But your confidence or resilience has deserted you, hiccupped as you put it, and you very nearly haven't survived ...'

'Yes, yeah, that's because I stopped being Captain Scarlet. I know I can't afford to let the Mysterons take over so that's my mission, to come up with a means of ensuring they never get me again. That's why I'm here isn't it? To work out how that happened, why, and to build a, to assemble a ... an infallible framework to use in the event of future attacks.'

She asked him about the voices he'd mentioned at the end of the last session. 'Were they Mysterons?'

'No, no. When I was really young I thought they were witches, they sounded threatening, and what they said always sounded menacing. There were three of them, women. Sometimes they would speak just once, and then I'd be struggling to work out what they'd actually said because they always spoke at the same time and in really low tones. But at other times they'd chatter at me almost continuously on the brink of my consciousness. It was tantalising, I desperately wanted to know what they were saying. I knew that it was wrong, I knew that I shouldn't hear voices, listen to people who weren't there, who didn't exist, but I didn't feel bad because no one else knew.'

'Did you tell anyone?'

'I told my mum once. She didn't seem very concerned, she told me to ask them to stop, or go away, something like that.'

The cats had struck up again but he continued, 'You see mental illness was never an issue in our house, because it simply didn't exist, still doesn't. It was like the pain thing, the not being allowed to cry thing, the never being allowed to miss school thing. Let's face it they'd been blown up in the war, literally in my mother's case, their fathers had been killed or died young, there was death all around them, so anything less was just not worth worrying about, just a trifle, just a flesh wound. So illness, or weakness, that just wasn't allowed.

'I don't think the voices were important. They were probably a reaction to stress, because I didn't have the loops or the seeing myself from outside things going on when I was a kid. And I know those are OCD things and I only get them badly when I'm really upset about stuff. Well, not the loops, they come and go all the time, but I'm lucky, they're short so I can live with them okay.'

'Loops?'

'Obsessive thoughts, they're like little film loops that run repeatedly in my mind, exactly the same motions. They last about a second and a half, and they repeat. Sometimes once or twice in a row a few seconds apart, at their worst about ten, maybe fifteen times a minute. But it's okay, they don't interrupt my life, they get a bit irritating I suppose but then because they are always exactly the same they are reassuring too, like they are a part of me and them happening reaffirms I'm still me. I used to try and control them, that's the compulsive bit, but that did make it worse so now I just go with them. I know … it sounds weird, but it's not as bad as it sounds.'

'What do you see when these little films run?'

'I get shot through my left shoulder blade and I begin to turn over that shoulder to see who did it. But it doesn't last long enough for me to see who it is. There's no sensation of pain, no jolt, it's purely a visual thing, I don't think I even feel myself

turning, I don't feel my collar on my neck or anything. I just know I've been shot and I'm looking back to see who has done it. It's been going on for about twelve, maybe thirteen years, always the same. Always exactly the same. And there's another one where I say "And that's what he does", it's me talking about myself, I'm the he.'

A diabolical howl pealed through the window accompanied by a low prolonged hiss that then escalated rapidly into a short vicious scream followed by the sound of claws scraping stone, a flowerpot thudding onto its side and grating as it rocked to a standstill.

3

Empire of Beauty

The Otter

May 1968

I SIT IN a peach-psychedelic papered kitchen-living room, with an otter asleep upon its back amongst the cushions on the floor, forepaws in the air, and with the expression of wide open bewilderment that very small babies wear in shock. On the tiled panel beneath the mantelpiece a board with the words 'Happy Birthday Chris – Seven Today' in watery poster paint and my mother's florid script. Beyond the door is the garden, whose fences back on the neighbours no more than a stone's throw distant, and encircling, mist-hung houses. A little group of house sparrows sweep past the window and alight upon the small carpet of green turf; but for the harsh, busy twitter of their voices and the sounds of the TV and the traffic there is relative quiet. This place has been my home now for seven years, and although I cannot even conceive of a future where I would ever leave I already know that I'll love this home until I die, this modest house with its three fellow human beings, not luxurious but comfortable, where my animals are welcomed and with its little garden where a necessary familiarity means the dirt beneath every broken half-brick and in the crooks of the apple tree show known and reassuring faces.

The otter was my present. I'd made it go to sleep under the sideboard beneath the window in the shadiest part of the room, the dimness offering a faint yet still desperately implausible suggestion that it might be real. Its body was cricked rather than curled, its tail snapped and its stiff straight limbs stuck up unnaturally into the shadow. They were blunt rods, footless, clawless and its torso was a tubby cylinder that fused necklessly with its rectangular flattened head. Two fraying ears flopped onto the cushion and its brown teddy bear eyes stared at me asymmetrically over its black button nose. Its fur was short, evenly coloured and glistened plastically, a sort of sparse mauvey-beige pile, nylon, through which I could see my mother's crude needlework.

Despite her efforts it was the whole world's least cuddly otter. It was worrisome, scary, like the handiwork of a deranged voodoo witch who wished to put a curse on the whole lutrine clan. It should be punctured with knitting needles in some maniacal midnight ceremony and cast by chanting hooded figures onto a pyre. It was a manifestation of otterised evil. But I loved it. And as the sky fanned out and gloom drowned the room, there in the clammy daze of the gas fire, as opportunity was knocking on the TV in the lounge, some animal magic worked for me and it transformed into Mijbil and I was married by delusion to my own ring of bright water.

I had been mad about dinosaurs but I was mad about bats when I got Maxwell's book. By the time I'd scrutinised the swatch of black and white photos I was mad about otters and by the time I'd struggled through half the text I was totally obsessed – everything went otter. Even my birthday cake had been 'otterised', my mother having baked several trays of sponge which she then skilfully carved and moulded into a writhing Mijbil. It was coated with melted Cadbury's Dairy Milk chocolate which she scraped with a fork to make furry, had orange Smarties for eyes and riding on its back were a row of toy plastic otters. Initially my infatuation was just focused on the animal, so I read all I could. First my encyclopaedias, then at the library, checking every index, turning to every page listed under 'otter'. But there weren't enough books, I chewed up Tarka the Otter immediately but thought it was childish, what I wanted was a textbook ... Otters of the World, pp450, with 1,000 full colour illustrations 2/6d, 'absolutely all you want to know about otters in one large inexpensive volume' ... but it didn't exist. And nor did the animal. Otters were extinct in Cleveland Road, Midanbury, in Southampton and Hampshire, probably the south of England. The local zoo was utterly otterless but by the time I'd finished Ring of Bright Water I'd written to London Zoo and they had kindly replied saying yes, they had otters. Bingo! Well, not quite because by then I'd learned about the man and his ottery aquabats, their relationship, his happiness, his passion, his tragedy, and just seeing an otter in a cage was never going to provide the antidote to my pathological fixation. You see I now loved otters as much as Gavin Maxwell did, so I wanted a pet otter too. And why not ... we had a bath.

The Dinner Table

June 1968

THE GREASY WET reek of bubble 'n' squeak crept around the banister on snakes of sour steam and slunk across his bedroom carpet fractionally pre-empting his mother screeching his name. He ignored her, listening for the chimes of The Magic Roundabout, knowing neither his sister nor father would move until Florence and Dougal had wobbled off screen and the hurdy-gurdy was rolling. The kitchen door was opened only so it could be slammed, the slurry of yesterday's vegetables was blackening, the re-reheated apple pie desiccating in the open grill.

Then the television popped off and a set of familiar creaks mapped their reluctant steps into the caustic fury of the dining room. The scowling hadn't mellowed as he sat down, placing his *How and Why Wonder Book of Dinosaurs* alongside him. He lined it up with the edge of the table, swapped the fork over to his right hand, placed it and the knife equally apart either side of the table mat, centred the dessert spoon and then lined up the salt, pepper and tomato sauce. He then reached out and set the four tumblers in an even grid. And then reset the book so that its top edge lay equidistant between these and the bottom of the mat just below the spoon.

The cover was fantastic, dominated by a ginormous fat Brontosaurus whose smooth grey neck arched to meet a surly creased face set with a hooded crimson eye, fixed in a very serious frown. Beyond the brilliant blue waters where this monster waded, an orange Tyrannosaur was menacing from a lush palm fringed jungle, and overhead a Rhamphorhynchus was fleeing a volcano spewing glowing lava and billowing a thick plume of smoke.

It was his favourite book, one of a series; he also had *Our Earth*, *Stars*, *Primitive Man*, *Prehistoric Mammals*, *Wild Animals* and *Birds* lined up on his bedroom shelf. He wished he had the whole set from the revolving display in W.H. Smiths, especially *Seashore* and *Lost Cities*, which he'd begged for but they'd said they couldn't afford.

She plonked a plate of cold chicken scraps, putty beans and fried mush on top of the prehistoric tableau, pallid and bleached and scorched. He didn't like Mondays, dinner was a misery, eating it was impossible. So he blew on the beans theatrically, muttered something about them being 'too hot', lifted the plate to one side and began.

'Tyrannosaurus. rex means "king of the tyrant lizards". It's got that long name so scientists all over the world know it's the same dinosaur. Scientists and children in France or Russia or Germany or South America all learn these names, like Brontosaurus which means "thunder lizard", Allosaurus which means "other lizard", Ankylosaurus which means "curved lizard", Protoceratops which means "first horn face", Triceratops means "three horned face", Stegosaurus means "cover lizard", Diplodocus means "double beam", Pteranodon means "toothed wing" ... Iguanodon means "iguana lizard tooth". They're all Latin and Greek words that tell something about that animal. Styracosaurus means "spiked lizard". So it doesn't matter if you speak a foreign language, everyone knows all the names, and absolutely everyone in the world knows Tyrannosaurus. rex. The rex bit means "king" and the Tyrannosaurus bit means "tyrant lizard" because a tyrant is a fierce ruler and Tyrannosaurus was the fiercest meat-eating dinosaur ever. And the most famous, more famous than Allosaurus which also ate meat but isn't as famous because Tyrannosaurus was bigger and a million times more ferocious. Because it had teeth which were six inches long and massive jaws and an absolutely enormous head and it was forty-five feet long and stood nineteen feet tall and wasn't afraid of any other living creature because its teeth could bite through all their skins put together and crush all their bones. That's why this Brontosaurus is hiding in the water, because Tyrannosaurus couldn't swim, but even so the Brontosaurus is scared, because it's heard the Tyrannosaurus roaring in the trees before it came out. But some dinosaurs didn't hide and they couldn't swim either because they had armour, like the knights of the round table only made of bone, which was heavy, so they would've drowned, they'd have sunk like a stone. So they avoided watery places and relied on their bony plates and spikes and their tails which had a huge lump of bone on the end.

'When Tyrannosaurus attacked an Ankylosaurus the Ankylosaurus would whack it with its tail and break some of its teeth and Tyrannosaurus would go and find something easier to kill. Not a Stegosaurus tho' because that was like a tank and had four massive spikes stuck out of its tail and massive bony diamonds all down its back, but it only had a brain as big as a walnut which could only work its jaws and front legs so it had another brain which worked its back legs and tail which it swiped at its enemies because it couldn't run because its back legs were bigger than its front ones because it ate plants on the ground. Triceratops did fight Tyrannosaurus and was generally quite a good fighter because it had a massive bony frill around its neck so Tyrannosaurus couldn't bite it and three massive horns on the front and it would charge at Tyrannosaurus like a rhinoceros, and the earth would shake and Triceratops would stab Tyrannosaurus in the stomach, but it would have to attack over again because it needed to get meat in its jaws so it would snap at the Triceratops' backbone until it had ripped off all its skin and torn off its legs so it could feast on its carcass even though it was injured itself. No one knows why all the dinosaurs died out, they were cold-blooded reptiles and when it wasn't warm any more they didn't like it and when all the plant-eater dinosaurs died there would have been nothing for Tyrannosaurus to eat any more so it would have died but this doesn't really explain the dying out because turtles, crocodiles, snakes and lizards were alive when the dinosaurs were alive and they're all still alive today so it doesn't make sense. So maybe little meat-eating dinosaurs ate the eggs and babies of the big meat-eating dinosaurs like Tyrannosaurus or maybe there was a disease which affected the reptiles but not the mammals which were already around, only small ones though, but then why didn't the germs kill all the reptiles like the turtles, the crocodiles, the snakes and lizards? That doesn't make any sense either. No one knows and that's really annoying.'

His mouth got bored and he stopped. They'd all finished and his sister asked if she could get down and his dad got up, sighed and walked out and his mum picked up his *How and Why Wonder Book of Dinosaurs*, dragged his plate back in front of him and went to do the washing up. He patted the skin of the Angel Delight with the back of his spoon until it went frothy. He couldn't find the book again before bedtime so he read about bats instead.

The Neighbours

June 1975

KEN PHILLIPS COULD still make out the wallpaper pattern through the staining, the criss-cross of blue and yellow bands that his young wife had chosen and they had hung haphazardly on that long-gone night when thunder thumped the window and lightning splashed gold into the feathers of her hair, as she stood like a doll dressed in his dad's dungarees, the legs rolled up, her little bare feet sticking to the offcuts of paper and her treading them from foot to foot as she tried to giggle them off. He'd flicked paste at her and then sloshed a brushload down her arm and in reply she had unbuttoned her shirt and invited him to stick himself to her chest.

She'd led him out into the storm and they'd rolled in the mud and grass until above their heaving and the crash of the rain she'd heard Tina crying. So she'd climbed off him and run inside and he'd lain there spinning, blinking up at the light watching the last strip of paper they'd hung through the window as it slowly peeled from the wall and curled towards the naked bulb where it shimmered before suddenly slipping to the floor.

When he'd finally staggered in and tracked her drips through the kitchen and up the naked wooden stairs the sobbing had stopped and when he nudged open their bedroom door they were both asleep in a tangled web of messy hair with the eiderdown pulled up to their chins. She was beautiful, he sighed and grinned, she was stunning, perfect. He'd had a fag and then re-hung that strip and finished the job as the sky greyed and the sun briefly silvered the big drips on the guttering and the birds were chirping and the milky clinked the bottles on the doorstep.

That was then; now the paper was peeling again, its edges deckled black, its yellow soured, the blue blistered, everything crusted with a thick coat of soot that thinned in patches to reveal cracked plaster and flaked paint, gobs of oil running from the scabs on the window frame, the carpet rolled back and heaped in a sodden lump up against the far wall. The kids had been wiping drawings onto the blotchy blackened door, a car, a plane, some faces and their names in block capitals all freckled with their smudged fingerprints in rosettes and rows of five and ten.

It had been three weeks since the fire and the whole house still stank even though the windows had been wide open every day. Every window except this one, the smashed one in Justin's bedroom, the one he'd nailed shut so the little rascal wouldn't climb out, couldn't climb out. Couldn't escape out of it after he'd lit a fire under his bed. He'd overheard one of the firemen calling him a 'fucking idiot' as he slumped there in the garden crying alongside the stinking mattress and flashing lights, Justin crying in his arms and Tina clinging to his leg in her shrivelled and singed nighty crying and her mother gone, gone for a year, gone except for Saturdays when she came back with her new hair and her skirts and her legs and her make-up, when she came to take them out for the afternoon, with her eyes and her goodbyes.

He rested the head of the claw hammer on the windowsill and stroked the glass with the side of his hand, scraping off a thin sludge of grot, which he wiped onto the charred frame. His kids were in the tree house, he could see them monkeying about with the three from next door up the hill. One of those Anderton boys had climbed higher, the other leaned at the back in his untucked school shirt, their sister and Tina and Justin were sat on the platform with their legs dangling. They all had ladles or spoons, Justin had their old red-handled potato masher and they were holding saucepans or frying pans.

Then suddenly they all looked over their shoulders up the road and when he leaned forward he saw his neighbour standing at the edge of her garden with a metal bucket resting on the only section of the fence that hadn't fallen over. Her waxy pinched face was talking to them and then they were all looking down the road, over into his other neighbours' garden. He watched her light a fag and drag on it as she hitched her pink sleeves over her flabby arms and stretched her thin top down over her belly. She parted her legs to bend down and dropped a rusty tyre spanner into the pail, then she said something to the kids and they all held up their hands, waved their utensils and cheered.

Down the road in the next garden, something ... it was the boy, Christopher, moving around in the shadows of the trees. He was wearing his camo jacket and slowly edging out onto the path with a brown bird perched on his left hand and as he knelt down it fluttered onto a small round block that was wedged between the paving slabs. He fiddled about beneath it, the bird stretched its wings and shook itself and Ken thought he could hear the tinkle of bells. He tapped the hammer, and whispered 'thanks'. He owed that kid, he'd sorted out their garden hose when the fire crew's hadn't worked, he'd leapt over the wall with it and back again to turn it on. Without it the whole bloody house would have gone up.

Meanwhile the tree gang were watching intently as Christopher turned away from the bird and backed up the garden trailing a length of string. Then he spun around, stretched out his hand and blew a whistle. Immediately there was an enormous eruption of clanging and banging, all the kids were drumming furiously on their pans and whooping and screaming, the boys jumping up and down, the whole tree shaking. Across the lawn his neighbour was beating her metal bucket, her fag pointing from lips drawn into a rancid smile. He looked back to see the bird fly off, pull up at the end of the string and tumble into some branches where it hung in a tangle of wings. A huge cheer went up, there was some shouting between the kids and the woman who then burst out laughing and lifted the bucket and spanner into the air in a gesture of triumph.

He turned away and started cleaning and after an hour of scrubbing and mopping he was filthy. He dropped his wet once-white T-shirt on the landing and padded downstairs with soot-stained knees and feet. After a vigorous bout of rubbing under the kitchen tap he managed to pinken his palms but his nails were clogged charcoal crescents, which he gnawed at whilst the gas woofed and the lard melted. He made their tea and called them in.

'Why were you banging that stuff up the tree for?'

Neither of them answered, they glanced at him and then each other and then returned to their plates. He looked at Justin, his elbows held high over the table, his grubby hands gripping two forks that were systematically attacking a flotilla of very crispy chips moored in a sea of tomato sauce that encircled a Wavy Line beefburger island. A shoal of peas had been beached around the edge of his plate and a thick spot of ketchup glistened on his chin. He smacked his mouth loudly and frowned as the burger sloshed back and forth refusing to be dismembered. Then he took a long slug of cola from his mug, wiped his face with the back of his hand and out popped a baby belch. He peeped through his flopped fringe and they both sniggered. The kettle was boiling, as the chair scraped across the lino he said, 'Didn't you hear me, I said, what were you doing up that bloody tree banging all that stuff for?'

He chucked a bag in his mug and was dazzled by a brilliant flash of light blazing through the open door; it was the pots lying on the dirt outside. The frying pan was face up and was being methodically licked by a crouched cat.

'And what are they bloody doing out-bloody-side?' he bellowed. 'Bloody hell.' Lunging towards the door he stamped hard on the floor and screamed 'Oiiiiiiieeee!' The moggy sprang up, faced him momentarily and then scarpered, bounding out into the long grass and slipping into a black hole beneath a shattered fence panel. He squeezed out the bag, dumped some sugar in, sniffed the milk bottle, winced, spat a 'Christ' under his breath and slumped heavily onto the chair, which skidded to a stop against the wall as the tea sloshed and scalded his fingers.

'The tree ... the banging ... what for ...?' he growled.

They looked at each other again and then Tina said, 'Missus

Anderton said to. It was her, not us.' She stuck her finger in her mouth to pick a chunk of chip from her teeth and then sucked it before stroking it quickly back and forth across her shorts. He glared at her and flared his eyebrows,

'She told us to make as much noise as we could every time he blew his whistle and as they had pans we got ours too. We've been doing it every day. Not yesterday when it was raining, but every other day. It's fun.' She looked at him. 'She told us to do it. It wasn't our idea. Honest to die.'

He sipped loudly at his black brew. 'And why would she want you all to do that?'

'I said, because of the whistle, because of him whistling.'

'Because of him whistling,' repeated Justin, nodding, the whole rubbery burger now impaled on his fork.

'The banging scares his bird, makes it fly away, like she said it would. Then he stops and doesn't whistle any more. That's all.'

'Jesus Christ, you lot made more noise than his bloody whistling.' He blew into the mug. 'Don't do it again, and bring the bloody pans in. And bloody wash them Tina, and bloody properly, do you understand me, Jesus Christ!' It was impossible without her here, everything had fallen apart.

The Bird

Sunday 27 July 1975

THE CITY TWINKLED prettily as the snug blanket of night slipped off its glittering bed; blocks of muted mauve, towers of indigo and long lines of lavender, soft and fuzzy, lazily drawing some colour from the eastern haze where the paling blue had already bleached to an albino blonde and had stolen its low-slung stars and washed the freckled face of the moon with milk. I could hear the buses revving in their depot five miles away and the clear chimes of the civic centre clock and watched a lorry speck stop and then counted thousands, waiting for the squeal of its brakes to race uphill, softened to a squeak by the nippy air and the murmuring vanity of the new day.

The tarmac and the dusty track from the bus stop had been swabbed with a dew that wet my thighs as I skipped through the mob of nettles, him on my fist, held high, fluttering fresh and new, chased there by the hum, squeak and whine of a milk float, the chinking bottles, the quiet whistle of the busy man and the clacking and clicking and swinging of a road of gates yawning, the bottles chattering in their crates, telling tales of yesterday's breakfasts and last night's cocoa and the tickle of a cat's whiskers on the edge of a bowl. And scuffing spray up onto the crest of the shimmering field I startled a fox, which gambolled away to the smudgy edge of brambletown, where it spun and sat, its coat smouldering like sunset, curling its brush with its torch-tip glowing in the bower, the drooping hoops of briars that arched from the high dome of the fortress where its cubs were peeping or already sleeping, rolling soft in the dry sweetness of their secret spaces.

I unhooked his block from my belt and planted it, stamping down the grass heads that annoyed his tail, tied him on and took out the creance from my gooey meat bag. He shuffled round, leaned forward and shook himself, and then scratched his chin with a rapid drilling of his bell. Satisfied that he was settled, I knelt down and placed the spring-clip that was tied to the end of the cord at the base of the block and then backed away from him, unfurling the thread that until now had held us infallibly together.

I laced it gently over the short turf that I had sneakily mown to prevent it snagging as it trailed behind him on his increasingly regular and longer flights to my fist. But this time it was just a ruse, I was only pretending, just going through the routine.

For the last seventeen mornings I had got up in the dark and snuck out to train him in those daybreaking hours, turning my back on insomniacs' cars to shield him from view, dodging the vampire dog walkers who ghosted through the bushes braying at their hounds. Wrapped in winter coats and wellingtons, I'd waited for them to crawl back to their coffins in the council flats, one after another in an irritating succession of slow ambling and aged plodding.

He'd been reluctant at first, perhaps bewildered by the scale of the world outside his garden – he'd sat statuesque, staring at me, bobbing his head, ducking and turning before weakly fluttering low above the ground and crashing onto my glove all fraught and frightful. And he'd swung off a few times, veering out and away until he was drawn up by the line, which then towed him into the turf in a fluster. I'd run over and crawled up to his furious gaping beak and crazy eyes, stretching out an apologetic hand with a generous ruby of beef, and when I'd picked him up it was all my fault. He'd leaned away all glarey, his dishevelled tail pressing on the back of my fist, his body wrung taut and skinny, and then thrown himself into a tantrum of angry bating.

And in these moments I saw everything that was pretty vanish and all the raw terror of a wild thing tied on a string, his trust torn by his tether, his tiny mind rioting in confusion, the deal broken amongst his panting panic and my tested temper as everything came undone and we one instantaneously became two strange species united only by a frail strand of nylon, strung in desperation to a destiny whose design could so easily disintegrate.

I was so scared by him, of him, of his prickly grip on me, his sharp sneers. I so wanted to make our world work, for everything to be perfect, immaculate, precise, controlled, for us to fuse in a place, an impenetrable, safe, secure and forever place, somewhere only we knew, some sort of fairytale Utopia, like Camusfearna, or *My Side of the Mountain* or as lone survivors in that strange post-apocalyptic wreckage that still haunted me from the shadows of Hiroshima and the three-minute warning. Just him and me and no one else.

When I'd laid out the line I walked back to him, knelt down and untied his leash. He studied my hands, stretching his neck to peer into my palms, he was keen. Then for the first time I slipped his jesses out of the swivel and let go of them. He was free. Untethered, untied. I stood up and walked away, my back to him whilst I fumbled in the plastic bag for a sliver of beef. I tucked it into my glove and kept treading my trampled path, I watched my wet shoes, tasting the metal whistle with my tongue tip, until I'd reached the ball of line.

I squashed the blob of meat on top of my thumb and went down on one trembling knee to ask the biggest question of my life. He was a jewel, radiant in the rich dawn light, his head bobbing. He was the centre point about which I danced, his tail fanning. He was all my absolute everything, his freedom terrifying. I lifted my arm and blew the whistle. He nodded. He shuffled. He ducked and flinched his drooping wings, he nearly toppled off the block and pirouetted as he steadied himself. He shook. I blew the whistle again, harder, freeing the sticky pea, feeling it whirling in the chamber.

And then he came, in the sweetest flurry of shallow flaps and ringing tinkles followed by a long low glide and a rush up onto my fist to crouch and snatch the titbit and flick it back and swallow it and I smoothly reached up and held the jesses and he was mine again and I breathed again. And the buses revved, the lorry pulled away and the moon spun and the whole world started over. And then I felt good. I'd never felt so good. I wanted to shout. I wanted to run, or jump, just run and jump. Instead I blushed, at him, and closed my eyes. I loved him so much I wanted to be him.

I floated back to the block, scanned for dogs, saw the fox slink into the thicket, felt the sun touch the gap between my soaked socks and trousers and lowered him through my shadow onto his perch. Again I released him and again he came back to me, four times in all, until I got too excited and knotted him onto his swivel and leash and wound it into my fingers. And I held him tight, and a dog bounded out, and I waited for the bin men to pass and I took him home and later that day all the queen ants poured out of the drain and I bought a Lolly Gobble Choc Bomb off the friendly old ice-cream man, which I sat and ate in his aviary and I teased his toes with the stick and I supposed that this was what they said happiness was.

The Naturalist

January 1971

HE'D FOUND A dead starling in the gutter by the bus stop. His mother had ordered him not to touch it but as soon as she'd been distracted by his misbehaving sister he'd slipped its light floppy body into his pocket. They went to the Triangle on a grand tour of the grocer's, butcher's and baker's and then traipsed back up

Thorold Hill and all the while he'd sneakily felt its sharp beak and barbed claws and stroked its smooth wings and pressed its squidgy eye. He couldn't wait to examine it.

When he eased the spangled corpse out of his coat after tea it was completely stiff and crooked with its head and tail twisted sideways and its wings clamped tight to its hard body. He had to wrench them open again and its pristine feathers became tatty and frayed.

His school compass's point was sharp enough to get between its beak, which he prised open so he could touch its tongue. The tiny slip of flesh didn't look like it could make any song, it was thin and rigid and wouldn't go back in the beak properly once he'd stretched it out with the bathroom tweezers. He teased back its eyelids too but its eyes had collapsed and shrunken deep into their sockets.

Finally he scrutinised its plumage. He was astonished by the sheer number of feathers and the precision in their alignment, the rows, tiers, layers that overlaid each other to form a seamless sequinned skin. He tweaked their dishevelled ends and edges and discovered that for all their weightless fragility they were eminently repairable. He'd soon preened them back to perfection and studied how they smoothly overlapped as the wings and tail opened and closed.

And the colours! A dazzling spectrum of shifting petrol-washed hues, purples, bottle greens and bronzed blues, they looked black in the shade of his palm but were electrified in the gleam of his table lamp. The patterns – a beautifully symmetrical fusion of spots and teardrops and arrow points somehow etched into the fabric, edged with golden or coppery lines that defined each plume and scaled the bird in flakes of miraculous fabric. It was so wonderfully made. He'd never been able to see a bird properly before and his gentle dissection, knelt beside his bed in the cramping cold, was a revelation that sowed the seed of a new obsession.

Paradise whydah: the name alone was exotic enough but the bird flew straight out of *Star Trek*, his flamboyant tail ribbons pulsing behind him as he flounced across Africa and pitched on a spindly twig alongside a dowdy sparrow-like female. She had been busy laying her eggs in other birds' nests, mainly pytilias, whatever they were. The male was two-thirds absurd tail, iridescent blue, maybe green, it was difficult to tell, but he had a brick-red bib that dripped down his cream chest and a short, dark bill a bit like a bullfinch.

But he was nothing compared to the nearby red bishop bird hopping through some luminous green vegetation searching for insects and seeds, probably to feed its young, which would be close by in its hooded nest, diligently woven from strips of reed blades and grass, the entrance near the top and slung like a cradle between the upright reed stems. In winter these black-faced scarlet beauties were dull, like the females he couldn't see, and moved around in flocks. It was cheeping, its bill was open, but there was no sound.

Swarms of tens of thousands of red-billed quelea swirled in the sky before funnelling down into the thorn trees that were filled with vast numbers of their nests, perhaps as many as ten million in just this one spot. They too had black faces but with crimson beaks, just like the violet-eared waxbill he saw hopping beneath them, a neat, slim, long-tailed inhabitant of the dry scrubland that laid three or four plain white eggs in its spherical brown grassy nest. He'd seen other waxbills in the pet shop; they were minuscule, nervous and nibbled on millet the same as his father's foul-tempered budgie.

The next bird he spotted was creeping over some moss on long yellow legs, and had a red and orange chest and a bold lemony stripe running from its blackbird-like bill to its bottle-green back. It looked a bit mean, fed on insects, termites, snails and slugs and hopped about on the forest floor where it was very difficult to find despite its felt-tip plumage. But for all its splendid colouration it wasn't the best bird; nor was the lilac-breasted roller even though he saw it swooping down to seize a juicy centipede on the sandy ground, fanning its spectacular wings and swallow-like tail. These gaudy hunters were quarrelsome and aggressive, especially during the nesting season when they attacked intruders, which the adjacent fish eagle certainly looked set to do, hunched on its perch frowning. He couldn't hear its loud yelping call but he studied it on its leafless branch over the shallow lake where it easily captured fish, especially if they were already dead or stranded.

Then he moved on to his favourite, the southern carmine bee-eater. From its high vantage point where it stood taut and keen it scanned for bees and other passing insects with its beady eye. It would swoop elegantly and expertly like an acrobat and dip down to its nest hole excavated in the sandy bank of the river. Four or five eggs are laid. Length of adult about fourteen inches. There was some glue smeared over its thin curved bill, which annoyed him, he wished his grandmother had been more careful when she had stuck the cards in.

He flicked a couple of pages to find the scarlet cock of the rock, which wasn't like any bird he'd seen or could even imagine. Its head didn't make sense, its beak was tiny and its eye wasn't in the middle, but in the painting on the cover where three males displayed in the jungle it looked stupendous. He'd love to see one but they came from Peru, which he'd found was pink on his globe and far away on the other side of South America. He'd been to France once, on the overnight ferry to Le Havre and back the following morning.

The text explained that as many as twenty males indulged in strange dances on the forest floor where one leapt and postured whilst the rest watched. He closed the page and studied the prancing orange marvels ... Brooke Bond Picture Cards, Tropical Birds, illustrated and described by C.F. Tunnicliffe. Price sixpence. Sixpence, fiddly and silvery, light and shiny. He flicked the torch off and slid the crisp card pamphlet out from under the eiderdown where he heard it flop onto the floor. He dreamed of exploring jungles and savannahs, those massive green areas in the atlas with no towns or roads, where new species of animals hid, where he'd find them, be the first ever human being to set eyes on their colourful wings, weird horns, fierce jaws, where he might even discover extinct reptiles ... dinosaurs. He would save all his sixpences from now on and one day have enough to fly to Peru.

He pulled down the covers and huffed a lino grey mushroom cloud into the screamingly cold air and then checked his snake light was still on. He could see Alpha, his North American chequered garter snake, basking on its slate plinth beneath the bulb, and alongside its vivarium the dark tank where Vite the green lizard had hidden until last week's escape. It was dead, he knew that, no cold-blooded animal could survive these temperatures, the flannel had been frozen to the taps this morning. He wondered where that lizard's body was, whether he'd ever find it and be able to get its skull.

He'd torn his bedroom apart searching for it, dragged every box out from beneath his bed, the wardrobe away from the wall, taken all the Easter eggs he'd received and stored for the last three years off their shelves, he'd even emptied out his old toy cupboard. There was no trace of the runaway but he had unearthed a dried seahorse, his best ever otter drawing and the two missiles from his Dinky Spectrum Pursuit Vehicle. He'd lost those again already though.

When his last garter snake died he'd tried to extract its entire skeleton by boiling it in a saucepan whilst his mother and sister were at ballet. It had taken ages for the water to heat up and because he'd had its rock-hard desiccated body hidden in his museum drawer for a couple of months it had just spun on the surface amongst the steaming bubbles. Then he'd gone to watch Grahame Dangerfield on *Blue Peter* and his dad had arrived home to find a stinking soup baked to the bottom of their best saucepan, which unbelievably he scraped down the toilet in a rage. What a waste, he was desperate for more skulls and snakes were his favourite animals at the moment. Cobras were the best.

The Snakes

July 1972

HE STOOD UP to check his bike was still leaning against the fence on the other side of the lake and a distant twinkle from the handlebars confirmed it was. Two swans whirred and wobbled low overhead and then crashed, spraying quick wakes of glassy spume, before they turned and stood up to shake and fold their candle-white wings, spinning to face each other whilst all the ducks muttered about their rudeness. He crossed his ankles and lowered himself down into the lush floppy herbage to play with his new pet.

As it slithered through the cage of his fingers he felt the broad flakes of its belly scales catching on his palm and listened to them click and rustle when it bent its body too tightly into a curve; he stroked its spine and chased all the little ridges on its surface down to its tail where the scales grew small and smooth and were tiled in pairs, criss-crossing like the braids of a pharaoh's beard and finishing with a sharp point that he pressed gently into his fingertip.

The snake was beginning to calm, it was still shoving with its soft blunt snout and dodging his hands when he passed it between them, but the thrashing urgency to escape was gone and he sensed its fear subsiding as he steered its looping neck into his lap. Its nose dug into the crack of his arm so he drew a soft noose of fingers up its body and slowly pulled it back, raising it up to his face.

Neat black lines inked its ivory lips and its bright beady eyes stared vacantly with an unfeeling intensity. He grinned; he adored grass snakes, they were his favourite things on earth, they were fast and wild, required a practised skill to find and catch and although each was subtly different in their patterning and tone of green, all were perfectly scaled, symmetrical and flawless – simply beautiful.

He wiped the residue of excrement from its body. It had vented this when he'd first snatched it out of the base of the willow and perfumed him with the pungent fluid, which was meant to deter predators. But it didn't deter him, he coveted that smell – it was a badge of success, it meant he had captured a snake and it would linger on him for a couple of days. At night he'd lie in bed waiting until the scent had faded from his nostrils before raising his fingers to sniff it again and again.

Below its ranks of ordered olive scales the 'grasser's actual skin was black and papery and when he twisted it over its underside was beautifully marbled in soapy blue-black and cream, each scale uniquely blotched and edged with a fine line of grey. Its polished jet tongue was still flicking furiously as it jabbed its head in search of refuge but as he massaged it gently between his hands its rate of waving steadied and soon the snake was curling lazily across his crossed legs.

He studied its breathing; it drew large sudden gulps that pumped down its body and were immediately expelled with a barely audible hiss. It was curious and it probed the world by arching its neck and tonguing the air as it steered itself through his embrace, unravelling a continual series of wafting explorations.

He examined its soft head and noticed its shape change as he gripped it firmly behind its lovely yellow collar. Its tongue fluttered at him, forked and shivering, quick and shiny and baring a brief smudge of pink at its root, and he drew its mouth up to his cheek, holding it still, waiting until he felt the little lashes sticky lick and then to the tip of his nose and finally to his closed eyelid where he felt the tendrils touch in a delicate spidery kiss. He wondered what the reptile sensed of him, whether he tasted or smelled good or bad, and when he'd unwound it and coaxed it carefully into his pillow-case snake-bag he rolled onto his stomach and wriggled through the grass, trying to sniff like a snake.

He sampled the busy nip of a crushed plant, the quiet earthy heart of a tussock and the noisy rotting tang of the water's edge. He squirmed on until he pushed his head out over the bank clear of the sedges and gazed over the midge-pimpled mirror out to the crowd of waxy lilies.

Their leaves flounced shamelessly to bare their fleshy under-sides, all but smothering their frilly white flowers as swallows parabola-ed and gilded their chests in the pond so close he could hear the little slaps over the distant sploshing of warring coots and the whirr of all the insect world. He stood up, peeled his muddied T-shirt off his belly, picked up the wriggling bag and continued to tiptoe around the clumps of marram and through the sword-leaves of iris that fringed the oasis.

Whilst chiff-chaffs chanted, he goose-fleshed in the cool of the alder shadows and scuffed the dew there to wash his ruined shoes, he watched fat blue dragonflies fighting, listened to the dry whisper of their wings as they clashed and parted, crisp and quick like the soft crush of melting snow, he startled a pair of ducks and smirked at them steaming away as their crèche of ducklings surfed and squeaked to catch them up, and glimpsed a kinky bauble flashing neon-blue before blackening on its b-line towards the stream, piping twice as it skimmed. It was Saturday morning, it was a kingfisher and it was great.

He kept the snakes in a large outdoor enclosure that he and his dad had made from six massive glass map plates they had fetched from the Ordnance Survey one evening in the spring. Each sheet was finely tapestried with a haywire filigree of numberless roads that knitted nameless towns to a grid of mysterious geographies, none of which he could ever identify. The plan was to construct a three-dimensional snake atlas but the panels were very heavy and they had struggled to heave them into the rectangular trench they had dug. There was also no budget for a supporting framework so to prevent the whole contraption from collapsing they'd wound rope around the top, tied in a tangled knot at one corner, which to his exasperation needed daily tightening to prevent gaps appearing in the seams.

In theory the sides were too high for the snakes to get out and for any mammalian predators to get in, and he could just about straddle over on his tiptoes to top up the enamel bowl with water and to arrange the rocks, logs and the large pile of elm bark beneath which his prizes liked to hide. But the snakes were superb escapologists and from the back room window he observed them stretching up towards the rim, gripping with their chins, quivering and slowly rising straight as sticks before finally toppling sideways and flopping into the thick clods of grass he had planted.

When he wasn't guarding them some of the lengthier captives made it out; he'd retrieved one magnificent specimen from next door's lawn where it was being lazily molested by their corpulent cat. It was unharmed but gone again by the time he got back from school the next day. But after a month of successful forays there were as many as six or seven grass snakes writhing in his vivarium and they needed feeding. Frogs would be an ideal meal but it was summer and they'd all vacated the pools and ponds; a trip to a New Forest stream had produced about thirty newts but these were gobbled up in less than an hour, mostly by one greedy 'grasser' that had coiled itself neatly around the bowl and systematically emptied it of the luckless amphibians. So tomorrow if all went to plan he was going to the 'minnowing spot' to get them a supply of fish. His mother and sister would have a picnic on the bank whilst the men waded out barefoot onto the gravel bank to chase the fawn fish and their chocolate shadows across the silt and through the weeds and onto the concrete ledges beneath the bridge where they were more easily cornered and scooped into their nets.

The Sparkle Jar

July 1972

IT FLASHED, IT swallowed the sunshine and sped it up and spat it out in beaming shards of super-shiny brilliance. Spinning glimmers and shafts of bright white light that darkened all the world around it and radiated flares, which blazed through the shadows and painted a rainbow that wobbled and glowed on the splashed paving. I gasped with joy and reeled in its shining, elated by my discovery of the thing I'd made, and I cupped it in my soft pink palms and I felt its cold and I drew my eyes closer to it to be dazzled. Everything seemed extra alive in that scintillating moment and as the gleams gyrated and glittered I imagined I could see their tiny twinkling hearts, seeding the sparks that made them so very vivid. And then I wiped away the spilled slop of the river, polished the glare and thrust my fingers into the sparkle jar to stir the soft tickles of the swirling tinsel of fishes.

The completeness, the excellence and absoluteness of that minute I spent teasing a jam jar of minnows and sticklebacks set on a post beside the stream, was immediate. Amongst the fresh bubbling froth of little fish, with their stripy sides and golden eyes, their gulping mouths and pumping gills and glistening scales, all dancing in the tides of my twirling fingers, I identified the essence of pure and simple and beautiful life. I was enraptured.

I caught more, delving my net into the gardens of weed, beneath the crumbling concrete that edged the footpath and around the flotsam of tyres and oil drums and pram and bike carcasses that had drowned in the brick-strewn shallows over which I stumbled in leaking wellies. The yellow nylon net filled with a soup of shrimps and silt so heavy it repeatedly fell from its bamboo pole and I had to wade beyond boot height to retrieve it, sloshing back to the shoal of gravel to sieve for booty, the greeny bogeys of fish sticking to my tingling fingers before I dipped them into the jar and they spun back to life with a determined wiggle.

Of course the bigger fry were further out, I could see them shifting across the honey-coloured mud in fast shoals, chased by their telltale pigeon-grey shadows, drifting beneath the sluggish wreaths and then vanishing downstream to dissipate in the reflections of the sky and the black water that sank into the tunnel beneath the road, where I imagined that I had once seen the spotted sides of a trout, for half a second.

Facing down the waterway I stood entranced by the beacon on its post at the edge of the wooded shore, the flaming jar still beaming starbright under the sycamores, still radiating life through its lens. Then over the gurgle of the drain and the road traffic voices came from the footpath, shouts and laughs. It was a gang of three and then five teenagers with feather cuts and Dr Martens, with Pumas and jeans and one of them had copper coloured platforms and they had Saints scarves tied to their wrists. My bike was laid on the grass and sure enough when they reached it one of them picked it up and sat on it, two of the others tipped it up and he nearly fell off. I waded downstream until they saw me.

'Oi nipper, this your bike?' one of them yelled, 'this your bike?'

I stopped opposite them and nodded. I felt the bricks slippery and the eddies riffling up my heels.

'D'you want it mush, eh, d'you wannit?' He took hold of the handlebars and demanded the kid astride it dismount. 'Darren gerroffit, gerroffit . . . '

The kid jumped off the saddle and sidled away, the one with the platforms started rocking it back and forth across the path, bumping the back wheel on the wall, then one of the others got hold of the seat hoop and then they counted down from five and on zero launched it out into the stream. It bounced on its front wheel and kicked up before somersaulting into a gritty crash of chrome and orange onto the reef of rubble. They whooped hysterically and one of them launched my rucksack in a high arc into a nettle bed on the other bank. I heard the spare jar smash.

'What'ya goin ta do about it?' they laughed. 'He's gonna tell his mummy and she's gonna give him a smack!'

Then they spotted the jar, which had been guttering in the dappling brown sway of the trees just as the sun caught it full on. It was set up like a coconut shy with a glittering prize for the winner. One of them bounded down to the shore and began tossing up stones to the others, who fought over the right to throw first. I began swashing out towards the far bank and the cement plinth crowned with the sparkling prize but the water grew deep and I had to stiffen my knees to press through it. I was about two yards short when I heard the whizz of the missile and simultaneously watched the flask explode with a cracking pop.

A little bang of luminous blue, a pulse of silver and a flopping thwack marked the end of my pocket universe – followed by a tinkle as the glass splintered around the base of the post. Thanks to that chance event the spectacular fission fed life to the water in the form of thirty or forty little fishes, which were already gone as I stumbled into a blackened hole and felt the shiver of the stream on my knees, my clammy sleeves stuck to my arms, the net pole juddering in the surge. All that life vanished in a single moment of violence and all that remained was a massive shocking vacuum that no one else on earth would ever experience or understand. Both the wing mirrors on my second-hand Chopper were smashed and the speedo was broken and there was a bright metal scratch like a Chinese character gouged into the chain guard. At home I watched *Animal Magic*. It was rubbish.

December 2003

'I tried talking. Every once in a while, maybe when a teacher asked me a question, but I'd normally only get halfway through before they told me it was enough, told me to sit down. It was the same with the other kids in my class. I remember at junior school, one Friday, near the end of term, there was no teacher for a whole lesson for some reason – that happened quite a lot then. Anyway, I had my dinosaur book with me and I tried to tell this kid, Duncan, about them but he kept interrupting, and then he shouted at me and walked off. In secondary school I spoke almost all the time, but I couldn't really talk, I tried a few times but they told me I was talking too fast, that it was boring, that I was a divvy. So I suppose in the end I gave up.

'Some of the other kids couldn't talk properly anyway, they misused the words they tried to use, one kid told me he'd been stunged by a wasp, no, no ... not stunged, he said it had stanged him. I'd sit there silently correcting him, his tenses were all over the place. His writing was a shambles too, letters back to front, words a mess. I realised years later he was almost certainly dyslexic. They called him a "thicko" and he was in the bottom set for everything and, you know what? There was probably nothing wrong with him at all.' He paused. 'I used to sit in lessons counting the verbal, grammatical mistakes the teachers made too. Once or twice I couldn't help myself correcting them, normally when they were telling me off, which got a laugh from

the rest of the class but got me stood outside or writing lines or detention.'

He appeared a little brighter this week, he was speaking more impulsively, he seemed bolder but was still not looking at her, still talking to a point that she judged to be the bottom of the right-hand leg of the table by the door.

'I did talk to my parents. And to my sister. I know they thought my obsessions were boring but at least they didn't tell me to shut up. Sometimes they'd start doing other things, I'd be telling them stuff over dinner and then they'd start doing the washing up and sometimes they would all just start laughing for no apparent reason. That was hurtful I suppose, although I can't remember it really upsetting me at the time. Sometimes I'd just talk to my dad, sometimes just to my mum, it would depend on whether I liked them or not at the time. If I didn't then I wouldn't say anything. I'd punish them, they couldn't have what I was thinking. I don't suppose they cared but it was my way of taking me away from them.'

'What about when you got older?' she asked.

'It changed when I got older,' he replied, 'in my teens ... then I learned to speak without saying anything. I could answer people if I had to. It was easier that way, it avoided conflict, and it seemed to work most of the time. Like I said, by then I'd pretty much given up on the idea of talking to people about the things which interested me or having actual, real, conversations.'

He shifted in the chair and smiled.

'My mum had bought me a small, blue paperback Collins Gem Dictionary *which I'd read and carried around all the time, I'd test myself on the notable dates in English history which were listed in the back. I've still got it.' He smiled again. 'Anyway, one*

afternoon, it was sunny, we were in the back room after lunch, my sister was Watching with Mother, I was blabbing on and she took the dictionary from me and wrote in very small neat handwriting into the front a poetic quote by Jonathan Swift ... "Conversation is but carving!, give no more to any guest, than he is able to digest" ... do you know it?'

She shook her head. He didn't acknowledge this either way before continuing.

'I could go on, I remember it, but the quotation ends with "Let thy neighbour carve for you." I think it was that which later made me realise that I had to listen rather than to talk. I still think of it now, all the time, every day, although I must admit that in truth by the time I remember to it's sometimes already too late. But I have a timer. That's how it works, I have a timer in my head. Only sometimes I don't hear it go off.'

She asked him what else he'd needed to teach himself. He mused for a moment and then said, 'I've learned to look at people I don't know. Actually, not so much learned, that's not right, I've trained myself to look at people. It was a conscious process. In nineteen eighty five.'

'But you only occasionally look at me and you know me ... '

Again he smiled. 'Yes, well sort of know you, but I don't know you enough. I can, could ... I do look at you. But I don't need to now. I presumably don't need to pretend to be what I'm not when I'm here and there's no one else here so I don't need to direct my conversation, I don't need to read you, to see what you're thinking, to see if you're lying, a, because I'm doing all the talking and, b, because I'm telling you the truth. That's the whole point of me being here isn't it, I've relinquished that control, I tell you the unadulterated truth and absolve myself of all control and concern of your judgement of it. And, I'm trying hard to relax, I thought that was part of it here too ... relaxing.'

Soon afterwards he left without looking at her.

4

Agitated Amazement Disorder

The Museum

August 1968

THE SUCKERS OF the octopus were rough to the touch but running his fingers over its eight winding tentacles felt good, the stone was dry and warm and with the tender palm of his hand he caressed the carved creature whilst gazing in amazement down into the shadowy hall packed with a jostling gang of gigantic skeletons.

This was a treasury of monsters, a time machine primed to transport his imagination back millions of years and enrich his dreams for years to come. Pressed against the terracotta arch he was still reeling from his first real-life encounter with a Diplodocus. The quiet gasp he'd spilled as he'd walked beneath its high outstretched neck and paused directly under its smiling skull had dissipated amid the sharp echoing whoops that caroused around the great hall, exploding from a swirling audience of other enthusiastic young admirers.

The space was vast, big enough for the largest dinosaur that had roamed the earth, the ceiling with its painted panels faded into a dusty haze of thinner air, the stained glass poured rosy shafts that bled all the way down to the wide plain of tiny floor tiles where they washed weak Ribena stains onto his socks and shins. He'd paced out the length of the creature, thirty of his sandal-slapping strides from the head, along the neck, past the great cage of ribs and down the steady descent of the tail to its final curled flourish on the plinth behind the barrier, and when he'd finished two more children were copying his idea, goose-stepping in straight lines whilst families parted in deference to their determined cause.

'Twenty-eight,' one shouted, to which the other boy exclaimed, 'Forty!'

He wound through a muttering cluster of adults hogging the best spot to look at the scale model of the dinosaur and slowly read the plaque. It said 100 feet, up to 30 tonnes and 150 million years old, Jurassic, a plant-eater found in North America. He smiled and his chest fluttered; it was, he concluded, the second best dinosaur ever. Better than Brontosaurus, Triceratops, even better than Steggy.

He liked the symmetry of this place; looking back to the entrance, he shuffled around until he'd arranged himself dead centrally and then gradually craned his head back. The roof was like the vast skeleton of a prehistoric reptile; an arcade of metal ribs held up the high ceiling and windows and led down to a series of mottled columns, between which yawned twelve chequered arches encrusted with vines alive with stone monkeys. An enormous staircase rose to a maze of corridors with countless enticing chambers and everywhere there were animals sculpted into the walls.

Despite its awesome volume, he felt immediately comfortable here. The dirty ochre of the masonry, the soft easy light, the cool dusty smell gave the museum a welcoming gentleness like a kindly old grandfather who, despite being important and serious, was sitting by the fire in the dark, chuckling and telling you all the world's most interesting stories in a whisper so they were yours and no one else's. But why were there octopus, flatfish, eels and cod all swimming up the pillars in the dinosaur hall?

His father marched ahead reading everything too fast, he could see his glistening Brylcreemed hair and white shirt flashing through the pools of light that lit all the brilliant things. When he first saw the Triceratops he thought it was a bit small; he stood back and compared it to the elephants he saw at the zoo, it was longer but only about the same height, no wonder it was always being attacked by Tyrannosaurus. But the skull was formidable, beaked and tusked, with that incredible bony frill, with massive eye sockets and a great big hole where its nostrils would have been.

He leaned forward; it had no smell but he peered into the crusty pots where its eyes once gleamed, and wondered. Beneath it was a fabulous model, about the size of a dog and so perfectly detailed. If he was ever rich he would buy that and have it in his bedroom. 30 feet long, 12 tonnes, 68 million years old, Cretaceous, again from North America. The third best dinosaur in the world ever. A bigger boy with glasses pushed him so he moved on. He hated being touched by strangers.

There were shiny stones the size of his fist that had been inside a dinosaur's stomach, Protoceratops eggs from the Gobi Desert, a Brontosaurus's left leg bone, coal black, gnarly and thicker and taller than him, and then there was 'Steggy', Stegosaurus. What a fossil! It was fantastic, the pointy skull was level with him, it was hunkered down with its toes splayed and its back arched, a mass of bronzy ribs wired to a frame and, alternately suspended along the spine, its flat diamond-shaped plates, starting with one about the size of his Chelsea F.C. pennant and rising up into the shadows where at the apex there was one as big as the coffee table at his gran's. It had the tail spikes too, just like his book, but they weren't very sharp.

He imagined this animal cowering beneath a ferocious predator, swishing that tail and slashing its flesh with those prongs, the deafening roar of dinosaur war, teeth shattering on the bony armour, blood dripping down its sides. 30 feet long, 5 tonnes, 155 million years old, Late Jurassic ... North America. Why were all these dinosaurs American? And why didn't it say about its main brain only being the size of a walnut and mention the extra one in its tail? If he met any experts from the museum he would ask them to add that onto the plaque, it was important information. And the model alongside it looked rubbish too, it was dirt brown, fat and grumpy with a down-turned mouth and its skin was all smooth, it had no scales. He had made a better one from a plaster of Paris kit he'd got for Christmas. He might swap it for that Triceratops.

His dad found him at the Iguanodon and started telling him things he already knew about it, but finding himself ignored the unpaleontological parent wandered off again back towards the 'Steg' where he stood in a beam of mustardy sunlight alongside the boy with long trousers and glasses. They both squinted at its plates for about twenty seconds and then left in different directions.

'Iggy' was 40 feet long, 4 tonnes, 150 million years old, Early Cretaceous, and had lived in Belgium, which was pink and next to Holland and nowhere near America. He looked down at its three thick toes and smooth blunt claws. It was a plant-eater, a herbivore as his book said, and the skeleton was pitch black and shiny.

He started to try and count the bones in its tail but couldn't, they were too complicated, with loads of blades sticking off them, the legs were all uneven and the ribs were wonky and the skull was so high up he couldn't examine it properly. And when he stood right under it and looked up towards the ceiling with all its posh swirly bits he could see the fossil wasn't symmetrical, it looked like it had been burned, had melted and its head was squashed.

He studied its huge hands, outstretched as if about to catch something, each with their characteristic thumb spike, which, his book said, the scientist who first discovered its skeleton had placed on its nose by mistake. If he stepped up onto the wooden plinth and stretched on tiptoe he might be able to reach those claws. He'd be gentle, he knew it was not allowed to be touched, and the museum was busy, that goggly-eyed kid was still mooching about and he was bound to tell and through the forest of ancient bones he could see a very glum-looking guard picking his nose at the end of the hall.

He bent down, untied his shoelace, scanned around, quickly retied it and then climbed onto the mahogany platform and reached up. The end of the Iguanodon's finger fitted perfectly into his hand, it was polished, like a big pebble, but not as hard and it wobbled a bit even though he wasn't squeezing it. He gawped at his hand and at the mighty black, pointed thumb just out of his reach … he'd definitely touch that when he got to the juniors. The moment seemed to last a long time, but it could have only been a second or so before he jumped down onto the wooden floor with a loud double clap. An old lady leaning by a pair of donkey-sized dinosaurs was smiling at him, she had bright eyes, twinkling, small red lips and perfect hair and when she lifted a single finger and touched it to her mouth to shush him he blushed and hurried away, even though he knew she wouldn't tell on him.

The Woods

July 1971

THE HOLE WAS cold. A fly zipped past him, he could taste the dirt, moist and fleshy, like in his granddad's greenhouse, he could feel the sun on his legs and a pricking pain from the scab on his knee. Digging in his toes he thrust himself further down until his shoulders scraped the sides and it went dark. When his eyes had adjusted there were wiggly roots and dead webs on the roof of the tunnel, the sides were smooth and the floor dusty. There were no clear footprints and when he breathed deeply through his nose he could smell nothing but the colour of the earth and when he held that breath and listened he heard only his heart and the ticking of the ground. Then his neck ached and he laid his head slowly down feeling a small twig gradually pressing hard into his cheek.

Straightening his arms and squeezing them to his sides he pushed again and slid further down than he intended. A sudden rush of panic made him wriggle. He could feel his weight pinning him into the vanishing burrow and now with dust in his eyes he could see nothing. He calmed himself and again laid his head on the ground, his ear pinched against his skull. He listened hard and waited. Maybe nothing lived there, or maybe it wasn't at home, or perhaps the tunnel ran much deeper into the hummock than he'd thought. He breathed out and it sounded loud but dead. What if he woke them up and they came to investigate, he'd never get out in time, they'd attack his face. One more listen … nothing.

It took him several anxious minutes to twist, pull and back himself out by bending and stretching and frantically clawing at the sand with his hands and toes and when he staggered up all filthy, panting and coughing, and began to shake himself off he judged that his feet had been in right up to the throat of the entrance. If he'd got stuck no one would ever have found him.

It was boiling and his sweaty arms and face were caked in grime and just as he was wiping thick smears of salty silt from his cheeks and chin onto his old school shirt he heard voices approaching around the path that skirted along the top of the knoll. Their tone told him they were bigger boys, probably from the estate, so he ducked down and bellyflopped into the thick shawl of bracken. Peering through the mesh of stalks he recognised them. He'd seen them before, out in the field smoking cigarettes. They'd been scary, with their jeans and belts with big buckles and Budgie jackets, and he'd turned back and run all the way round through the woods to avoid them. Now he froze as they lurched past and then sloped off towards the monkey swing. It was a close one.

Rolling over, he found he had dirt grinding in his teeth so he swilled some gob around his mouth and then spat it out, a string of dribble swaying down to darken a spot between his legs. Sat over the hole he idly trod footprints in a row across the sand, his worn plastic soles methodically marking ovals in a regular pattern. Then he had a genius idea ... if he smoothed all this sand flat then he'd be able to see the prints of any animals that climbed out or crept into the hole.

He knelt into the entrance and began palming the loose soil, picking out all the twigs and leaves as he worked his way back up the slope onto the mound of earth that had been excavated from the entrance. But then he realised it was only just after lunchtime; loads of things could run over the surface before it got dark and that would ruin everything. He'd have to come out again later in the evening, he'd probably have to sneak out of his window but it would be worth it to find the tracks of a fox or a badger. Then he could use some plaster of Paris to cast them like he'd seen on *Blue Peter* and then start his own collection of different footprints.

Maybe one day he'd get an otter, that would be his real prize, an otter footprint for his nature museum. This archive of broken shells, dishevelled feathers, a couple of tatty birds' nests, his fossils, some bones, skulls, a sticky owl pellet, a pair of starling's wings and some dried fox poo was carefully preserved and meticulously labelled in his top drawer. All his clothes were now stuffed into the bottom two, the other having been converted to a hangar for his expanding squadron of Airfix model aircraft.

The wood was overslept, drugged and lazy, its darknesses rich, black and cool, slashed by scattered shafts of sunlight that cut through the bleached palette of listless green and in a few spots burned firework-bright on the ground. There were no birds singing, not a chirp, just the pulsing hum of insect noise and a baby crying in one of the flats behind him and then the distant chimes of an ice-cream van and then the Fords factory siren.

From his perch he could survey the slope all the way down to the torpid stream, where old pram and bike wheels were fossilising in the petroly pools of rainbowed water, where when you trod through the ginger bog the muck beneath was black and bitter and its nip never left your shoes, where he'd slipped on a swamp-swallowed sheet of iron and gashed his palm and it had throbbed as it oozed dusky ribbons of blood, which tickled his wrist on the run home and his dad forced his hand under the kitchen tap and the water from the cold chrome bit hard and he wasn't allowed to cry or get stitches like everyone else.

He jigged back into the juicy surf of bracken that jungled the area around the badger holes, following their paths edged with a pulped fringe of bluebell leaves, and slipped and slid as he crawled up into the steep shade of the chestnut tree, the sticky juice of the spent plants glistening on his toes and knees, stringy between his fingers.

Squatting on his log, he picked at the crispy scab on his knee; there was sand under it so he scraped it with his nail and blew. The grains were stuck fast so he pulled at the crust and teased it back to reveal a glossy blister of fresh pink tissue – raw, internal, made of something leaking out of him. He tried to glue the scab back down but it wouldn't go so he pushed it hard with his thumb to see how much it hurt ... not much, just a twinge and then numbness. Then he gently peeled it off and rolled it in his fingers until it crumbled, whilst blowing tickles of cool air onto the wet dirt-ringed wound. When he puckered up the skin a tiny amount of blood seeped out of the tingle, which he dabbed onto his finger and licked off.

Suddenly a magpie swooped up into the canopy right above his head, ricocheted off the branch in terror and rattled away over the shimmering surface of the ferns. It had made him jump and he forgot about his graze and followed it into the woods. The track twisted like a vein and was soft and silent as he trod over the older boys' prints and wound his legs around the triffidy tendrils of bramble until he reached a drowsy clearing.

They rose twinkling in a cosmos of beautiful light, spinning and flickering, flashing their brown wings white, flaring brilliantly in a helix of spangling twists and loops and then fusing briefly before pulsing apart and rising again, livid and lucent, flaming furiously, spiralling above the tangle of briar and bracken. And then there were three, four, sparking in a gyre of soft and silent fury.

He'd seen them before, sunbathing, spreading their cream-spotted wings flat on leafy platforms along the path side; sometimes they would burst into the air and reel away as he ran through the woods leaving a swirling confetti of startled insects in his wake. And he'd crept on his knees and watched them snapping their wings shut, revealing the plain chocolate wash that radiated from their hairy black bodies across the speckled slivers of tissue that flinched and finally spread leaving the glowing silver V of their antennae pointing at their tiny tufted heads. And then, when they were set they'd sit, their legs tightening them down in robotic jerks, snapping them into a statuesque stillness on their pedestals where they would blissfully bake in the full blast of the sun.

But now they were fighting, a mob were wrestling and writhing in a super-fast scuffle and although it had begun with three or four, from nowhere others had flown into the radiant fray and there were six, seven, eight, they were hard to count such was the frenzy.

He sidled closer until they were right in front of him and oblivious as he ducked into the shadow of the bank and rose with his face into the circus of light where they tumbled. He could hear them! The quick rustle of their tiny wings when they smashed together and squinting skyward he could see quick puffs of dust exploding as they collided after arcing out of the ruckus and crashing back into the splendid crucible of jumbled wings.

And then they all collapsed down and fluttered onto his face, he felt the papery kiss of their bodies brushing his cheeks as he held his breath and closed his eyes listening to the fragile riot and when he sensed them rising he watched them shimmering up the column of sunshine, high into the canopy, sparkling like stars and then vanishing into the roof of the wood, one or two darting between the leaves, black flecks against the faultless blue.

He was on fire. His skin burning, he slumped back onto the coiled roots that knuckled the bank and wiped his face, rubbing his eyes until a galaxy of luminous spots raced across them, the ghosts of the butterflies re-formed into a kaleidoscope of rushing neons and fluxing mauves chasing blobs of turquoise left to right as he pressed hard and tried to squeeze all the dazzle out so he could see again, and when he peeled open his eyes and squinneyed out into the woods everything was bleached and bleary, the shadows milky and grey and the great wave of bracken a frothy haze of lime-washed foam floating on a scaffold of inky blue sticks.

Without the restless insects the place seemed stunned, stupefied, shocked by that ballet of gossamer violence, the wonder of plain and simple things drawn together to conjure such beauty, transforming that bubble of urban air into a theatre where an astonishing performance was fleetingly played to an awed audience of one, the memory of which would sparkle for a lifetime. And he knew it then, in that moment of dazed happiness, what a gift, what a thing he had seen, what a treasure he held.

He sprinted home across the wasteland, his feet clapping a rhythm on the hard-baked path through the fallen fringe of dried grass that nibbled his legs. He leapt the trickle that leaked from the spring and then pounded up the slope to the ridge where a tatty flotsam of pigeons flaked low and applauded through an accelerating arc to crowd under the sun, their shadows speeding over the track.

He paused to spin round and embrace his kingdom, consciously drawing all the elements of his territory together – the great blonde plain rugged with bushy islands, the valley with its prickly pillows of brambles and the thick wooded eiderdown snuggling up to the shoulders of the bluff, capped with the flats of the estate. Through the stewed air he saw strata of trees and copses and fields receding far beyond the lines of rooftops and the muzzy sides of the van factory; one day he'd explore there too.

This was his paradise, bursting with so many exciting things, and as he turned and smacked his soles down on the pavement, zigzagging in staggers and bounds to avoid stepping on the cracks, skipping along the kerb by dropping his left foot into the gutter, dodging dog shit and curling round the lamp posts, then putting his head down to race up the rise to his gate, which he flung open so hard that it bounced on its hinges and clattered closed behind him, he thought that he would never leave here.

And then as he panted up the sideway he heard them shouting in the kitchen, he smelled the frying pan and listened to the vicious clash of metal as the cutlery drawer was rammed shut, his father stomping to lay the table between each of her terse commands. He wasn't allowed to talk about his butterflies at teatime. No one talked at all.

The Dinosaur Film

July 1969

THE POSTER FOR *One Million Years B.C.* was magnificent. I couldn't get it out of my head, it was totally amazing. But my horrible parents didn't think so. Hidden on the stairs I eavesdropped on them moaning about it whilst they were doing the washing up. He said it was 'less to do with dinosaurs and more about barefaced sex' and scoffed at the bit that described it as 'a savage world whose only law was lust!' She echoed his disgust with something scathing about 'mankind's first bikini' and then got really hissy about 'that legs akimbo pose'. I had no idea what they were on about but those were the reasons they wouldn't let me see it. They didn't understand; of course I was interested in what was going on between Raquel's legs – it was a caveman seeing off an Allosaurus with a spear.

Every night throughout December I pawed over the cinema listings in the local paper. It was there, at the Regent, with afternoon showings, certificate A and I begged them. I told them I didn't want anything else for Christmas, I didn't want the big Scalextric set from Edwin Jones, I just wanted to see that film. It would save them a fortune, my sister didn't need to go, just me and Dad on Saturday instead of going to Grannie's, she wouldn't mind, I'd tidy my room forever, get the coal in for the fire every morning, go to bed early, clean my own shoes, wear a tie, be nice to my Aunty Ruby. In truth I'd have eaten myself alive to watch that film but they wouldn't have it. My mother had seen 'that starlet' with her legs spread and my dad was familiar with her Jurassic cleavage and I was too young. I was also utterly distraught and thus I hated them. For three years.

To try and placate me, at the end of January they took me to see *Ring of Bright Water*. I didn't really like the way the film ignored the real story and I hated the bits with Mr Maxwell and the woman who wasn't even in the book. And the comedy sequences with the otter. Otters weren't funny – they were serious animals. But nevertheless I loved it. I cried at the end, got told to 'grow up' by my father who then refused to look at or talk to me all the way home. In the foyer he'd paused to light a cigarette and complained to my mother that 'it was bloody embarrassing'. But she bought me a record of the theme tune, sung by her Saturday night favourite, Val Doonican. I loved that too and insisted it was played at full volume on the downstairs gramophone every night, at least five or six times, whilst I lay in bed preparing to dream about my solitary life in the security of a wilderness with only animals as friends. And my dad didn't approve of that either, Val was 'soft' and the song was 'sentimental rubbish'.

Movies would play a run and then came back to town again so I became obsessed with the front page of the *Southern Evening*

Echo's Classifieds section. Each Thursday the left-hand column advertised the titles that would be showing that week.

There were several cinemas in Southampton: there was the Gaumont, which only showed a few films as it was really a theatre, and was where I had endured the interminable *Dr Zhivago*. At least I'd been allowed a Kia-Ora in the intermission, a rare treat, they were usually deemed 'too expensive'. Then there was the Odeon, where I'd had to cope with my mother smiling inanely in the flickering glow of the abysmal *Sound of Music*, twice, the second time whilst she mouthed the lyrics out of time. The Woolston Picture Theatre where, through an apt but impenetrable fug of cigarette smoke, my father and I had glimpsed the fall of the Alamo – again. The Atherley in Shirley, with its glassy brassy ostentatious foyer, the scruffy Classic on the high street and the Regal at Eastleigh where I was 'treated' to *The Jungle Book* because it was 'my sort of film, all about animals'. I loathed it. The cartoons were unrealistic and it had singing all the time.

Lastly there was the Regent, which we had never been to. It advertised intriguing titles such as *Bang Bang*, *Casting Call*, *Girls That Do* and *Come One, Come All*. Some seemed to have an 'animally' theme, *Faster Pussycat, Kill! Kill!* and *Vixen* and one entitled *Sex in Sweden*. I asked what that might be about but didn't get an answer, so I enquired again. The paper was snatched off the table and I got shouted at and then, sulking in my room with the big dictionary, I got no further: 'sexual intercourse', 'reproductive functions' and 'fornication' ... all meant nothing. So I got out my dinosaur books and revised the geological timescale so I'd get it right when my dad next asked me to recite the order or test me on whether the Tri came before the Jurassic, which it did.

After just such an epoch, on a Thursday night in July my dream finally came true. It was on again! At the newly renamed ABC in a double bill with another film called *She* and it was still an A, which meant I could get in with an adult! *One Million Years B.C., One Million Years B.C., One Million Years B.C.*. . . I'd probably bleated it one million times even before my dad got home and then a million more before bedtime.

I remember them being in a really foul mood that evening. I don't think it was my infernal nagging, my fossil collection strategically laid out on the sideboard alongside my neat stack of dinosaur books or my best ever drawing of a T. rex pinned to my bedroom door, I think it was them again. It was every weekend now without fail, the grunting that became screaming, her chucking the crockery on the table and him walking out halfway through dinner. Then the kitchen door slamming and the muffled roaring from the front room, then her storming back and ordering us to clear our plates, smiling, talking about something nice in a childish tone, pretending for us, in that voice, too cheery, too soft, like a lullaby in a firestorm. Then we'd be sent to our rooms and I'd draw some mushroom clouds and lie awake waiting for the three-minute warning and my sister would undress and re-dress her dolls in the colossal silence that haunted the aftermath of their cold wars. But somewhere amongst all that menace they agreed that my dad would take me to see it on Saturday afternoon. I lay there in my bed mapping the streetlights' pattern on the ceiling, not beside myself with excitement, but terrified that something would go wrong. And it did.

I hid in my bedroom all morning, only sneaking down to check the showing time, twice, two thirty, two thirty. I didn't want to risk annoying my mum. She was in the back room with my sister sorting out the button box, doing some needlework. Then they nipped up to my nan's and to the corner shop, Austen's Stores, to get some revolting packet soup and some vegetables. I snuck into my sister's room to watch them leave.

It was unusually cold; we had no central heating or double glazing, just the gas fire in the front and the coal in the back. My room had a two-bar electric heater we'd got from my uncle that I dropped things on through the grille to watch them burn, plastic was best.

He got home before them and shouted up to me because the doors were unlocked. Then I heard him make a cup of coffee and sit down to read the paper. They were ages, they were late, and when I finally heard her bashing the pans about I knew it wasn't going to be the usual Saturday 'sandwich lunch'. Eventually she growled my name, I sighed, stood up in my scrupulously tidied cell, padded down the stairs rather than slithering down the banister, opened the kitchen door and through a billowing gush of bitter smog turned quietly into the back room and sat down at the table.

There must have been twenty, some the size of eggs, most the size of large marbles. They covered about a third of my plate in two layers, pallid greyish-green and faintly steaming, piled around a shiny pink circle of gammon and a small dollop of almost mashed potato. I immediately reached for the tomato sauce but she cut me off. 'I'll do it,' she snarled, 'there's hardly any left.'

She shook and slapped the bottle but only a watery splodge the size of a penny blotted the meat. I asked her to put some vinegar in the bottle but she said we didn't have any. The only other thing she said that lunchtime was that if I didn't eat all my sprouts I wouldn't be going to see the film. I cut a few up, spread them about and when my dad left tried to eat them. I gagged and spat them back onto my plate. At about three thirty she came in and took the plate away and told me to go to my room. I hated them. I didn't cry, I could have, should have. But instead I drew a highly detailed map of an imaginary island, with bays and beaches, lakes, forests, mountains. One house, no roads, no people. Just me and the animals.

She took me to see it the following Saturday before he got home from work, left him a note and no lunch. And after all that it was ahistorical nonsense: living animals, a green iguana and a tarantula, mixed in with giant turtles that never existed, dinosaur species romping together when in my books it said they had lived millions of years apart and cavemen in the Tri-Jurassica-Cretaceous when they were only first around two hundred thousand years ago. And there weren't enough scenes with the fat and pimply dinosaurs and they didn't move anything like real reptiles. But Raquel Welch was utterly sensational, her tyrannosauric bosoms squashed into that flimsy doeskin bra had special effects and I fell massively, madly and secretly in love with her. And with Ursula Andress from the other film. She played an evil queen but when she entered one scene in a tantalisingly translucent gown tied with only thin braid I had to stifle a 'wow' – she was beautiful. When I grew up I'd marry one of them and we'd never eat sprouts.

The Bird

Sunday 17 August 1975

HE ROSE OFF my fist and turned up tightly, tinkling and fluttering in a distinctly flappy flight that I thought appeared playful, but then just as a drop of rain slopped on my cheek and stuck like a big cold tear he fanned his tail and for the first time gave a clearly agitated cry. Then he shrieked and sped straight out across the field climbing very fast.

Something was wrong. My heart heaved, I put the whistle in my mouth and as I began to run I heard another call, the excited whickering of a second Kestrel. It was high, roughriding the churning sky on flat wings, black and small, rotating out, getting smaller high over the red roofs of the estate as I spun looking for him. There was a pause where I heard the pall of rain rising in pitch, felt the smack of it on my shoulders and then blinking into the barrage I watched him converge on the stranger, listened helplessly as they whinnied, watched as their sinuous spirals merged, until they canted on a tilting circumference and balanced across a diameter with a shrinking radius.

I stood as they shrank, specked amongst the warring rags of cloud far out where the sky was falling like a hammer, over the factories, over places he didn't know, weaving away in that vast volume I couldn't reach, and then they connected with that point where all little things finally vanish. He was gone.

I ran, ran through the weir of blasting rain, my new Green Flashes soaked in the thick grass, squelching as they pounded the fresh black tarmac of the estate pavements, slipping on the road rainbows. I ran between the houses, through the cut ways over the greens, under the willows, panting and pounding, to the gravel track that led out to the river.

I could taste the running blood in my spittle, the stitch bit hard in my stomach, I spat and staggered out of the trees, trying to find all the sky I could see, jamming my breath to listen and hearing my heart thump, the whistle wheezing between my teeth. He was gone. I stood terrified in the dead light as my world quaked and a tonne of hopelessness fell, crushing my chest with a choking sadness, and the spectre of loss, of a finality, of the impossible, suddenly became shockingly tangible. That he was gone forever, that my dream of a perfect life was already over.

I began jogging again, along the riverside, pumping on the whistle, screaming his name through the deaf trees up into the dumb sucking hole above me, I panted to a beaten standstill and wished I wasn't ground-bound, that I could fly, there was just so much more sky than land. Then I reached the road through the mill and dodged recklessly through the cars and staggered into the lane that wound out to Fords. He'd never been here, he was lost, wandering in countless dimensions, all invisible, all inaccessible to me. I listened through the traffic, ran out to the roundabout and on to the row of allotments that edged the airfield. I was miles from home now, he'd been gone for an hour.

Desperate, I turned around and began backtracking, skipping the puddles, scanning the rooftops, shouting and shaking spit from the overblown whistle. By the time I'd reached the far side of the field the fists of cloud were finally cracking apart and a sharp light bled under the roll of treacherous rage that had spent its rain.

My legs ached and the stitch was permanent as I trudged up the slope onto the crest to the worn patch where the hole for his block was and slowly turned, scanning every visible part of the dripping watercolour around me. I could see the places where I'd been searching, tiny trees and toy-town houses in neat lines, the shining roof of the van works. But that space stretched out infinitely all around and over me and its vast scale thumped me with an enormous emptiness. He was gone. Stolen away by an innate attraction to his own kind. It said something.

I hung there, watching the odd person pass, with their dog or pushchair or partner, a father with his boy uselessly trying to get a kite into the limp air, him running, dragging the juddering blue plastic at head height, the boy towing it over the grass until it tore and they went home in tears. My mind raved as the turmoil in the sky drained away and the sun sparkled from a field of diamonds sprinkled over a billion blades of grass and, squinting at a volley of happy house martins chirping and buzzing like flies high over the shoulder of the copse, I tried to turn every distant dove, pigeon, every glinting crow into a Kestrel through the misted lenses of my binoculars. Then I limped home.

They were all out but he was not preening on the roof of his aviary as I'd fantasised. Back in the field it was as if the storm had never happened, it was warm and drying out and in the dip between the fox brambles and the sandpit a gang of kids from the estate were trying to have a picnic, arguing over Smoky Bacon and Corona and jamboree bags and jelly snakes and the kite man returned for one more abortive attempt.

I sat cross-legged and traced vapour trails across the sky with my fingertip from places I'd never been to places I'd never go. I analysed and reanalysed that lunchtime, critically unpacking the order of events and their eventualities, finding a hundred ways to blame myself for what had happened, all the things I did or didn't do that led to this horrific mistake. I projected forward, a day, a week, years and wondered how I'd ever reconcile my guilt and cope with the loss, what I would now do before school, at lunchtimes, after tea. And I agonised over what would become of him – would his jesses become entangled on a branch leaving him to dangle in a lonely treetop gibbet, to mummify in some secret place, food for flies? Maybe he'd swoop down to a stranger, mistaking them for me, and then starve in their shed as they fed him bread? Or would he be shot by some gamekeeper and be tossed into a ditch to sink beneath the silt and moulder in a broth of stinking shrimps?

Suddenly a boisterous mongrel was lolloping around me in a fit of manic wagging and yapping and then sniffing at the meat in my bag. I jumped up and snatched it up onto my shoulder. It leapt, bounced, its back legs kicking up, white-eyed, its gums bared and drooling as I turned in a circle showing it my back. When its bawling owner arrived it fell off me with its tail down and slunk away, cowering and trotting well ahead of him back to the path. And then I heard his bells. Just once, distantly. My heart burst and I exploded into a panic. I could tell by the short refrain that he was perched, that he'd shaken or scratched.

Grabbing my binos I spun on the spot and anxiously began to scan all the treetops. Nothing... maybe I'd imagined it. I began striding down the field towards the woods, through the chaotic demise of the picnic, stopping to pan across the thick cushion of alder and oak, quickly but methodically scouring the mosaic of shadows and the shimmering jigsaw of greenery. Then I heard it again, far off behind me; this time he was flying a short distance.

By the time I'd legged it back up the slope I could hear the constant peal of his bells but it was coming from further up the hill, back towards my house. I began swinging the lure and nearly blew the pea out of the whistle, my teeth chattering as it went berserk in its tiny chromed chamber.

He was there, weaving amongst the roofs, half-heartedly pursuing starlings – so rather than running the long way round, I hurdled someone's fence, sank momentarily in their soggy compost heap, fell forward as the air erupted with wasps, before racing the enraged jaspers up the sideway and out of their front gate into the road.

Following the jingling I tore around the corner to see the little hooligan settle on our neighbour Mr Phillips's gable. He shuffled, I stopped, he bobbed his head and then let out a sharp excited trill before stamping his feet and mantling. He recognised me. But he was cross and he was hungry. I lifted the whistle to my lips. I needed to blow it. But what if that Anderton cow from next door came out? She would scare him off. Her car was parked beside me, she was in.

I took a breath ... and then blew hard. And again, and again. It took a few tense minutes to coax him down between the tree and the web of telephone wires and he refused to land in the road, eventually tumbling onto the tablecloth of grass and dragging the lure beneath Tina's toppled toy pushchair and the face-down body of a naked doll. I crawled nervously forward, stretched slowly and then snatched hold of his jesses and for a few seconds it felt like the whole National Grid was surging through my body and my giddy head.

She was standing on her doorstep. But she wasn't looking at me, she was frowning down the hill and when I knelt up Phillips was stood there. He was pale, shirtless, his jeans were undone and his hair was all sleepwrecked and one of his toes was poking out of its manky sock. He was pointing a finger at her. His hand was shaking slightly and when I heard her door slam he slowly lowered it and faced me. But he didn't speak or smile or anything, he just looked through us for a moment and then turned and went in.

The Dream

August 1975

THE SKY WAS still dark, a big blue velvet bay beached with amber streetlights and their gleamings from inky windows and the yellowed sheens of rain-washed cars and the Christmassy glow from a freshly snow-cemmed semi. Everything was static. He dropped the curtains and analysed the luminous dots on his alarm clock, ten to three. Then he must have drifted back off.

His chest lightened as he banked up hard, the air riffling his feathers, tickling his legs, he squeezed his toes together and heard his wing tips whizzing as he belted into a big curve out of the shade into the dazzling sun, slipping across the sky so fast it made his nose run. And then into an instant white-out, a shocking cold on his eyes, tearing tiny tears from the mist, racing in little rivulets over his back and down his tail and through the fluxing honey glow of the fading sun all strewn over with a tracery of fine threads, then the world flashing far beneath him as he hurled out into a vast canyon of puffy grey vapour. He tensed his shoulders and finned through the roiling eddies of invisibility, into the darkening heart of the cloud, chilled and dulled, silent except for his speed, until he flashed out and over a huge golden sheet that stretched away to where the world curved and the great slicks of sky converged into a thick charcoal ribbon. He watched the sun race him over the sea, he saw strings of twinkling lights snaking away and all the little things moving in their world far below, he saw other birds beneath him, drifting backwards and disappearing under the ruffling muslin of mists and he knew everything behind him was gone.

Again he felt the wind tugging at his wings so he tightened and began to rise fast away from the fading earth, up in a giant circle, through shades of orange and red towards a vault of deepening blue, and so he rose until he could see nothing but the landscape of the air stretched out forever, and below, floating in a skein of the faintest fibre, the most spectacular ridges and ranges of towering wraiths rose in the smudged form of slowly moving monsters, immense poodles, recumbent Teddy boys, dismembered dragons and maps of Australia, Surrey, of melted Wales and in the shape of that birthmark on his cousin's neck. Their sheer sides held valleys and caverns big enough to hide giants whilst their flat plateaus and fearless peaks reached vainly towards the stars pinpricking the void beyond the vapourlands, the flawless sheet beyond the sky webbed only with thin wisps of silvered light. He heeled down and played amongst the frilled edges of silky ghosts strung on a slip of turquoise veined with plumes rising from a bonfire of bruised lilac to another sky, veiled with a gilded web and torn tufts glistening as they floated free and faded away in front of every shade of shadow. And finally, as the dying light rubbed fire onto the last of the big tops high up in thinner air, he closed his body and fell towards the earth, cleaving the darkening strata, the guttering sun flaring through the last lightning-lit seams of cloud, the air screaming with the glory of his gravity, he sank into the night, into the warm murmur and fizz of their world, from a place where he was one, to the confusion and chaos in the realm of millions, spiralling down, fearlessly racing the perspective of everything expanding superfast, confident in his dimension and then instantly terrified of the crash into theirs, where rather than alone he was lonely.

The White Spider

June 1976

CARELESS RED BEARDS frayed the pepper-freckled glass and fuzzed the four corners entirely. She had daubed the frame of the mirror that hung alongside the kitchen sink in scarlet gloss to match the shelves, which teetered under a bric-a-brac of crockery, almost masking the thrice-emulsioned wallpaper that had bubbled up and peeled behind it all. We washed and cleaned our teeth in the kitchen, filling the plastic washing-up bowl from the kettle unless the water was on, which it only was when we were having a bath and there were enough coins for the meter.

He'd gone. His father always drove home for lunch. He'd make a sandwich, usually Cheddar with a couple of pickled onions, and eat it with a slice of fruit cake sat in his chair reading the *Daily Mail* from the front backwards. Then he'd go to the toilet, blow his nose loudly and drive back to work. In all his father had been at home for just under an hour and throughout that time he'd been sat on his bed listening to him moving around downstairs, trying to read, nervous. He'd lain awake late into the previous night planning and replanning what would happen next.

He removed his 'school clothes', not a uniform, but what they let him wear, and opened his wardrobe. Smarting in a gust of naphthalene he eased out his navy blue Brutus shirt. It was too small now but was the only fashionable one he had, with large floppy button-down collars and a matching cravat that he discarded onto the bed. Next he took out a mint-coloured pair of forty-four-inch Oxford bags with a high waist and two thin belts. He had bullied his mother into buying these in the Plummers sale. She had bluntly resisted all afternoon as they had criss-crossed the high street from shop to shop, but eventually he'd nagged her back to the rail and in exasperation she'd given in. 'Jesus Christ!' his father had exclaimed, and then minced across the room squeaking 'who's a nancy boy then' before blowing him a kiss.

He put them on and tucked in the shirt, its buttons straining, and hitched up the knees before bending down to reach under his bed to pull out a new beige shoebox. With it arranged on his lap he removed the lid and parted the tissue paper. Burnished oxblood, spatulate, with fake leather layered heels and six brass holes with thick ribbon laces; he smelled them, plastic and new, size 8½.

He'd bought them in Tru-form's with his paper round money, withdrawn from the Post Office the previous Saturday, and he'd shown no one. It wasn't worth it. He fingered his heels into them, tied the bows and stood up, lifting the curtains of his trousers to admire them. They looked massive, awesome, exactly the same as Lee Fitzpatrick and Kenny Parker had. As he plodded around the room they thwacked the floor loudly and the sides cut hard into his ankles so he pried them off, rolled on a second pair of socks and then tottered carefully downstairs leaning forward, hanging hand over hand on the banister.

He opened the downstairs toilet door and found his coffee-brown skinny-rib zip-up turtleneck jumper in the wash basket, sniffed it, put it on, did it up so you couldn't see the stretched shirt buttons and turned to face the mirror.

His nose was massive and his ears stuck out. He reached up through his fringed bob and pulled them back so they disappeared under his hair but when he let go two prominent crescents of pink poked through again. Then he pushed them from behind, looked dead ahead and pouted. He looked like a chimpanzee; his lips were massive too and protruded. Twisting his head sideways he peered out of the corner of his eye and studied the relative size of his nose and lips, tipping his chin up and down and pushing against both with his fingers, deploring their fleshiness.

He slipped off his shoes and sprinted upstairs to his mother's bedroom and spent ten minutes using her dressing table's side mirrors to evaluate his face from all angles, then picked up her Silvikrin and dashed back to the kitchen. Using a comb he wet his hair and made a centre parting, spreading his long fringe into two feathered wings either side of his forehead that he blasted with generous puffs of hairspray and held in place until they had set. Finally he trickled the last dregs from the little green bottle of Brut he'd swapped off his cousin down his clammy chest before he locked the back door and waddled down the sideway.

This is what would happen ... He'd finished all his exams so he didn't need to go back for any lessons. But he would walk down past the school as everyone else left at going-home time. There would be a handful of fourth- and fifth-years who had been doing their O levels and CSEs and all of the second- and third-year pupils. She was in the third year. He would be walking past her somewhere on the hundred-yard stretch of Little Dell Road, on the right-hand side. She would be with her friend and wearing a white blouse with her black cardigan, with her handbag over her right shoulder, her hair tied back, black trousers with her white ankle socks and black sandals. She was never one of the first out of the gate bang on 3.45, but not the last either, he would have to time it perfectly, approximately 3.58 to 4.04.

He would loiter at the bus stop in Woodmill Lane and when he judged that enough kids had passed him he'd amble down and round the corner and she'd be there. And she would see him and look at him for the first time, and she would smile, he would say 'Hi', she would say 'Hi', her friend would disappear and then he would ask 'What are you up to?' and she would reply 'Just walking home' and she would smile again and they would wander back to Spinney Walk and by the time they'd got to the cut through from Old Farm Drive he'd have asked her out, and she would nod and say 'yes' and on Saturday they'd go to the cinema in the afternoon.

The streets were empty, it was scorching hot, tablets of toasted grass edged with baked earth and dried flowers, burned marigolds and roasted geraniums, everyone had their windows and doors open, their nets hooked behind window levers, he saw shadowy figures sweltering in kitchens, abandoned toys scattered in driveways, this would be the day the earth caught fire, he was already sweating. There was no shade at the bus stop and he stood fluffing the collar of his cardigan, wishing he could take it off.

A new white number 4A crawled laboriously up the hill and lurched to a standstill, the doors hissed open and the driver's silhouette leered at him, propped against the timetable, expecting him to get on. Instead he walked away and didn't lean against the wall and squinted into the bus. After a contrived hiatus the guy turned, shook his head, looked down at his wheel, mopped his brow with his wrist, the doors juddered closed and after another exhausted pause the bus pulled slowly away just as the first lad sprinted up the pavement and staggered up to its closed doors. It was nearly time.

There were a lot more kids about than he had expected, hopping off the kerbs, shouting, kicking each other, ganging up and rampaging. He dodged through their mayhem and reached the halfway point. He was walking too fast, slow down, slow down he told himself. One or two made eye contact with him and he glanced away quickly but then he reached the corner, he could see the gates, people milling about, some teachers' cars emerging, he was dripping. He stopped, turned, hesitated, then dashed back to the red brick wall by the garages, he would wait here until she turned that corner.

Flyaway threads of bindweed curled hopefully into the sky from the top of the wall, which was bushed with a lush fragrant shroud studded with soft white trumpets. He backed into the vegetation, stumbling on some bottles tangled in the dead dog-pissed grass. The insides of the flowers were veined with a soft silverlight, gently grading to a cool lime at the base of the cluster of anthers, which was capped with a fuzzy sherbet of pollen. He plucked one, pinched off the end of the cone and held it up to look out through it, getting a plant's eye view of the road, the kids passing, the garden opposite with its shady grove of cherry trees.

He imagined a bee approaching, bumbling and busy, flying into his eye. Then he peered into another flower and to his surprise found one of those lovely marble-white spiders that hide there waiting for their prey. They were so superbly camouflaged that he'd only ever found one before, when he'd been checking a hidey-hole in the fence around the electricity station next to the Castle pub to see if he could find a missing porno mag. He couldn't, but in the creeper that clogged the chain-link he'd spied one of these beauties and taken it home. It had died – he'd put it in a jar that still reeked of pickled onions and that had poisoned it. He'd felt really bad about it.

Snipping off one of the plants' winding tendrils he inserted it into the flower. The spider flexed its forelimbs open aggressively and he could just make out bracelets of microscopic black bristles around their joints, its minute fangs and a cluster of tiny eyes on its forehead. Again he teased its legs and it jerked sideways clawing at the trembling sliver of green, and when he irritated it for a third time it slipped off its perch and tumbled out of the flower into his cupped hand where it balled up momentarily on its back. He moved to cup his hands but in doing so stimulated it to flip over and scurry rapidly up and over his fingers and fall onto the pavement below. Again it paused, before edging slowly but steadily out into the sun, towards the gutter.

He peeped round the foliage up to the corner. There was no sign of her but a bawdy cluster of second-years was cavorting towards him. He looked down; the spider was struggling out onto the searing tarmac wasteland, it would be trodden on, and it would be his fault, and he'd killed the last one, and it was so beautiful.

He stepped out and turned his back to the oncoming herd, sheltering the spider. He'd have to be quick, she'd be there any second. But when he presented the runaway with his palm and tried to chivvy it on, it repeatedly refused, turning sideways and scurrying around the edge of his fingers. A drip of sweat formed on the end of his nose and the kids parted around him, becoming silent as they passed. Two glanced back briefly; he looked over his shoulder panicking. His shoes were killing him, he'd have to kneel down, but his Oxford bags ... he knelt and began scrabbling after the spider. But he just couldn't coax it onto his hand. He squinted over his shoulder. She was there, just turning, with her black cardigan draped over her bag, with her white blouse on, black trousers, socks, sandals, smiling at her friend.

Flinching, he turned back to the spider and tried to gently pinch it up with his fingers, but it had scuttled into a patch of coarse asphalt, got wedged amongst some small tarry stones.

She was laughing, they were coming straight towards him, twenty yards . . .

The tip of his pinky was smeared with tar, he flicked at the sticky gravel. One of the spider's front legs came off and it stopped moving, he mopped his brow, he could feel sweat cascading down his scalp, running behind his ears, down his neck, his fringe was stuck to his forehead, his heart pounding, the vulgar niff of after-shave. She was about ten yards away, they were going to trip right over him. Using his nail he dislodged the stone chip, picked the spider up, dropped it into his palm, fisted his hand and stood up. Right into her face.

The moment lasted forever. Probably only three, four, or five seconds became a hellish eternity. He was taller than both of them and by jumping up he had surprised or scared them. They were open-mouthed, offended. She saw him, she looked at him and she began to frown. He was frozen, he didn't think to smile, he later relived that part of the incident repeatedly in front of the mirror trying to see how he had actually looked, how bad it really was. He had a million things in his mind but couldn't use any of them, he was paralysed, terrified, distraught, ruined and blushing ferociously. She sort of tutted, her friend passed behind him and they were staring at him, they were scowling at him, then they turned away and whispered and then they glanced back giggling.

His head was ablaze, he felt faint, stranded, how would he get home, they could all see him, everyone was looking. He looked down, his knees were smudged with black bitumen, the toes of his shoes had deep tan gouges cut through the red polish, his face was fluorescing, he was on fire, he could taste his salty sweat, he opened his wet fist, the spider was smashed, just a greasy blob, he shook it off and it stuck to his thigh, he flicked it onto the ground and began walking fast, head down, not running but almost and when he passed opposite the bus stop three fifth-years jeered and called him a 'fucking bender'.

April 2004

'As a student, at university? Well, it wasn't great because I hadn't begun to get a handle on my self-management. I think I still saw myself as that born-broken savage. I didn't interact much, tried not to get involved, I didn't fit in so I didn't mix in. But it didn't get to me like it did at the end in school. I remember my eighteenth birthday being shit. I got a bus out to the New Forest and wandered around all day in the cold on my own. That was rubbish I suppose. I remember hurting that day, being resentful, envious maybe, but in my heart I knew it was too late. And by then it just didn't matter as much any more. Because ultimately even if I had wanted to fit in, I couldn't, I'd lost touch, I couldn't even speak to them. I was isolated, but I wasn't lonely, there's a difference, you need to feel alone amongst people to be lonely, to care about the barriers between you and them. They'd gone so I was free, relieved, excited.'

The rain was gusting in sharp flurries of shot against the glass behind him and the room was dim and smelled of stale wetness, his clothes, his hat, shoes, trousers all dripping. His face was in shadow, he was unusually unshaven, he looked pallid, ill, fragile.

'Were you shy at university?' she asked as he stroked his face, pulling his hands down his cheeks, wiping his sopping hair back before sitting forward, his back off the chair. He was more animated this week, not crushing himself into the chair in an attempt to merge into its form and fabric.

'I don't think so. Aren't shy people frightened of social situations? I was never outwardly fearful of that, except with girls, I was very shy anywhere near girls. Petrified. But no, I just preferred being on my own. As soon as I'd learned what they were about I just couldn't get interested in their lives, I couldn't summon the energy for it, it was exhausting listening to them talking inane shit about ... them. As it turned out we had nothing much in common and really, I suppose, I was only interested in what was going on in my life ... or more pertinently, what was going on in my head. They all spoke louder than me, they'd all be blabbering and I knew I couldn't have possibly got the right type of word in. And when very occasionally I did speak they weren't interested anyway, in the end I suppose they thought I was a mute freak. If I hadn't looked like a freak they might have thought I was aloof or something, but whatever, it wasn't worth fighting it. I just got on with my work, I pretty much just worked and that was it, what else could I do? And I loved it, learning, studying my badgers, shrews, all that stuff.

'It's not that I didn't like a few people, the people who were close to me, those I could trust, or at least thought I could...' He paused. 'But by the time I was sixteen, seventeen, I'd realised that I couldn't trust anyone superficially. They were all in constant competitive conflict with each other. As far as I could make out none of them genuinely liked each other. It was a horrible shabby scramble, everyone clawing and biting over nothing to gain nothing and telling each other all about it. Teenage? Hell-age more like.'

He snorted softly, shook his head and went on, 'And do you know what, it seemed to make them happy. They were always happier than me, or at least appeared to be, told each other they were. They seemed to genuinely like themselves, liked each other liking themselves, it was all such a big "wow", such a laugh, such a good time, they thought everything was so good, they seemed so blindly optimistic.'

Now he sat back, pushed himself into the frame, tucked his arms into those of the chair, aligned his hands, his fingers, sighed and said, 'They were only ever looking out, they never looked in.'

'And you despised them for this?'

'Despise them? Did I despise them? Not really, I just couldn't relate to them and any negativity was motivated by my envy. Occasionally I wished I could be happy like them, I'd wish I could join in and that made me angry, it made me angry with myself. I tried to internalise it though, to reduce outright conflict but underneath there was always resentment. And not now of course, I don't now, not at all,' he said brightly. 'It wasn't their fault' he shifted slightly, 'you see I know the problem was really mine all along, in that I thought they had a choice. Back then I thought they were too cowardly to think deeply about themselves, that they chose to protect themselves by loving themselves and their world, but now I know they had no more choice than I did, we're just wired differently, different parts of our brain are a bit more developed than the others. They're out, I'm in ... it's the way it is.'

'So have you forgiven them?'

He thought for a while and then said, 'No one at university needed forgiving, they didn't do anything wrong. But the kids at school ... no.' He paused and shrank a little. 'I should, but I haven't, of course not. I can, could if I wanted to, relive all those nights, all those days, those words, looks, sneers and whispers. They're all still there.' He crossed his legs and arms simultaneously. 'I haven't got over that by forgetting ... just by remembering less often.'

5

My Piece of Sky

The Egg

April 1973

THE JACKDAWS BLEW off the field like bonfire ash, freckling grey into the bone-white sky, breaking down to the roofs, catching up on the chimneys, their chatter spent in the gaunt wind that heaved up the slope and leaned into him like a shoving crowd of ghosts. Torn grass, paper, crumpled cans and a flurry of polystyrene edged the sticky clay path checked with prints from Clarks, Millwards and the Co-op and in his pocket he fiddled with Peter Osgood, broken off at the knees, a casualty of last night's big match on his bedroom carpet where he'd beaten himself at Subbuteo.

Snow carrion from the sheet that had fallen on the first Saturday afternoon of the Easter holidays lay cowering in the shadows, irregular volumes flecked with dripped dirt, hard, with the taste of spoons, they popped with a kick and then dribbled down his cuffs as he revolved them in his bright pink fingers searching for symmetry.

He was out for the sake of it, the front room was choked with smoke, his sister was going to a party, his parents had rowed over the reheated roast and he'd flushed his sprouts down the toilet once they'd got bored with nagging him and ignored his retching. He should have put his cousin's old boots on, his white plimmers were now rimmed with mud, which would scab them for weeks and stain them forever. On the way down the hill the reckless buffeting had felt good, only he was out in spring's winter, he was the brave one, but now brawling with the gale had become tiring, his head was down, his zips were up, he stooped into the hedge and settled on the rusty oil drum that all last summer he'd hoped would hide lizards but had only ever sheltered slugs.

From his cave he could see no one or anything man-made and the squall blanked out all their noises too. A sudden pelt of hail walloped him and he curled into the tree and closed his eyes savouring the brief polished fragrance of new ice that pinged with the bouncing beads over the ground. They could all be body-snatched, they could all die from a lethal plague, he'd be the sole survivor, the omega boy, he'd fortify a den, get guns and fight the zombie hordes and he'd have all the animals he wanted for company and he'd have purple carpet and those modern inflatable chairs in every room, he'd dress like Ed Straker and he'd never eat another sprout.

The bullets of hail blasted through the flimsy drapery of leaves and popped across his lap, lodged in the folds of his sleeves and thwacked the plastic bag that kept the can company in this rustic refuge, walled with bramble, holly and hawthorn. He huddled in thought, in fantasies of survival for what seemed an age. He didn't know hours or minutes, his watchless time was measured only by his hunger and tiredness, not by his parents' curfews or even the passage of the sun. He roamed his world hourlessly – but now he was cold and fed-up, he'd go home, to his room and put the transfers on his freshly assembled BF109E, maybe start on the 262 if he could squeeze enough glue out of the twisted tubes, but mend Osgood first, he'd need him for the cup final that evening.

As he stooped out into the bleak, something brisk and brown tumbled through the rocking twigs and twitched and vanished into the mossy wattle at the foot of the bank. It was gloomy and thick down there … a mouse? But then a flicker, and then a little bird, sharp-billed and gone again through the ground as shy and swift as a lizard. He squatted till his ankles hurt and then as he stood he caught it out again, pinballing through the wind-swinging tangle and scuttling away into the undergrowth.

He crouched and waited. Nothing – so he slid his finger down his sock and rubbed his eczema spot, which was wet and itchy, he tasted it, chewed a scab, and then noticed a clot in the drooping wicker of the holly, hooked up where the briar webbed the bushy basket together. It was round, with a flatter top, a bit spiny at the edges but curiously capped with fresh greenery. It had a definite structure, it had been made, but he couldn't reach it, the cage of thorns was too dense.

The cur was on its hind legs, up and flailing with a skidding skinhead in tow, all canines and creased flesh, a borstal cartoon with a raw-nippled belly, high-leg oxbloods and Harrington and Ben Sherman all lurching and lunging to the snaps and snarls of panic and chaos. He'd startled it as he'd dashed out of the bush, and the cobalt-scalped bloke now grappled with the chain that was barely preventing him from having his face chewed off. The beast charged, choked, and flipped stiff-legged onto its back, shiny gloves hauled it over the mud, coughing … it spun, and its master bellowed, 'DAAMMMAAAGGGE!!!'

But Damage was done for the moment, dragged off, waddling and sneezing towards the councils. The skinhead pulled his jacket down, turned, called him a 'stupid little cunt', and then on their six bowed legs the couple jiggled comically down the hail-stung scarp like hamstrung puppets whilst imagining they strode like gladiators. He watched them vanish and then turned to the bush.

Blue. But like no blue he'd ever seen, five lovely eggs, pristine, clean, smooth, not Chelsea or Everton blue, more Man City, but richer, thicker. He trembled and scanned furtively around the field, then pulled the spiky leaves further apart with his cuffs and lowered his face into the foliage. It was so neat, the little cup, so wonderfully crafted from such simple things. His eyes raced over the circular cushion of emerald moss ringed with an intricate crown of grass and twigs that knitted so seamlessly into the mesh of thorns.

And sunk into it, a bowl spun with a lace of hair, fine, and smooth and grey, black, white and a few strands of reddish brown, all harvested from dogs that had snagged them whilst they paused to piss along the hedge.

But the real treasures were the five jewels that glowed so brightly in the greylight that bleached from the chalky sky. They radiated a colour so clear, so unique, that he was in immediate and absolute awe of them. He glanced nervously over his shoulder at the barren stubble between the houses, blown empty by the ruthless gusting. A bus was just pulling away, nudging the neck of the hill, and two flapping macs were trudging face down along the slope towards him, wind-bent and drooping garish Fine Fare carriers crammed with things. He folded the green window shut and sloped away until they'd passed and then ran back to the nest, pried open the curtains and they were even bluer.

He imagined they were incomparably fragile, he'd been holding his breath over the cup but now he was actually reaching out to touch them, steadying his wrist on the prickles as the tree tossed in the blustery scuffle. He could barely feel the shells but he hooked one of the eggs into the crook of his finger and rolled it into his palm. It was warm! Little warm like the baby mice he'd held when they'd nested in the shed, once, before they'd all disappeared. He turned his back to the wind and shielded it with his hands cupped close to his chest and tried to judge its negligible weight and tried to see the blue of it hard enough so he'd know it forever, when he closed his eyes and thought of it, when he was grown up, when he was old. Then he put the egg back, rearranged its secrecy and went home.

The Observer's Book of Birds' Eggs had a simple cover, white, with three specked and two blue eggs, neither of which were like his; they both had spots, his were unmarked. The far right-hand aisle of Wisemans bookshop was his unofficial library. It was narrow, cluttered, and the tall shelves cast a conspiratorial shadow on his regular visits, lots of furtive reading – but few purchases. The ivory postcard-sized pages revealed two at a time a catalogue of life-sized eggs and descriptions of the nest and the time of year it was built. But there were quite a few bluish eggs, even plain blue ones. He separated the pages with the most likely candidates between the fingers of both his hands and then turned between them cross-referencing the site of the nest, its fabric, the habitat and of course the illustrations of the eggs – none of which matched his colour of blue.

It was dark by the time the tall bald man had whispered to the short grey lady to go and tell him they were closing and the wobbly door shut behind him, jingling its bell. He walked up the Broadway, past the stack of buses outside Woolworths where washed-out women lined up at the tills in the stale yellow light – sunken dreamboats smelling of cabbage and cigarettes in scarves and scuffed shoes, fumbling in their purses, dragging their kids by their hoods, past the dark bike shop with its glimmering rows of chrome handlebars and past the florist's where they were slopping buckets of stinking flower-stewed water into the corner drain, to the bus stop outside Kelsall's where a long queue for the 4 and 4A meant he'd have to stand all the way to Selma Court, the new houses they'd put smack bang on top of his favourite ants' nest.

His mum had said it was okay to take one but that the mother might smell him and desert the nest. His dad had told him that he would have to blow the egg or it would rot and stink his bedroom out. *The Observer's Book of Birds' Eggs*, twelfth edition, 50p, told him that the egg resting on a tuft of cotton wool on a saucer in front of him was a hedge sparrow, Prunella. modularis. He took the needle and twisted it to bore a small hole in the broader top of the shell and then turned it over and pricked another at the other, pointier end. They were watching Z-Cars in colour, he was scared stiff of breaking the most important thing he had ever owned.

On the Monday it had been too dark to go back to the nest, on Tuesday there were people walking their dogs but he'd sneakily felt the eggs and they were cold, on Wednesday and Thursday he'd seen the mother fly off a couple of times. Then he'd got the book with some money from his gran, some cotton wool to line an old Oxo tin from the shed, which he'd tipped his dad's best drills out of, and then he'd become an egg collector.

He licked his lips and very gently blew at the narrow end, taking care to hold the egg over the safety of the rolled-up wodge of toilet paper he'd put on the saucer on his bed, him kneeling alongside it. He blew again and this time a minute pearl of clear fluid appeared and he wiped it onto the tissue. Again he blew, a little more liquid, he blew harder but nothing more leaked out so he put it down.

It was no longer as blue as when he'd first found it. Then it had seemed to gleam with its own vivid energy, to be magically generating its own internal light, to be luminous, glowing through the fine shell to radiate a unique new tint of the spectrum that was clearly beyond the palette of artists and couldn't be reproduced with the pigments available to printers. Now it held a cloud within the shell, its shade had congealed to produce an uncomfortable intensity, an almost noxious, sickeningly thick azure.

Rested and calmed, he made the holes a little larger and began to blow again but only ejected a tiny viscous tear from the egg. Maybe he wasn't blowing hard enough. So, shifting his grip to delicately pinch the capsule between his thumb and the pink of his fingertips and pointing into his palm, he braced his wrist on his chest, craned down and pursed his lips over the frail blue globe. He blew and the egg collapsed.

The crumpled shell was soft and stuck to his fingers, the exploded mess glued to his hand, which he gradually opened. He felt sick with the shock of it, he had the gummy taste of raw egg in his mouth and now knew why he had been unable to force its contents out of the pinhole. Looking like a bubblegum bogey bathed in shiny spittle a fully developed sparrow embryo lay on its back – bulbous-bellied, big-headed and black-eyed, with a broad waxy bill and peg-like wings and legs, its toes pricked with minuscule claws. He turned it over. It was dead. If he hadn't stolen the egg it would have hatched by now, or at least by tomorrow.

The following afternoon he returned to the nest hoping to find it full of the other chicks. But it was empty – no eggs, no young. It was warm, one of those days spring sometimes borrows from summer, and there were Persil-white clouds gambolling over the sky. He loitered on the path near the bush, dumbfounded, confused, angry, and saw the field beyond dapple dramatically into shadow. He watched the dark edge wipe rapidly across the ground and then, strangely, slow just in front of him – he stood lit but the rest of the world had gone out.

The towering cloud unfurled its edge, shoved the sun's face away and spewed a vile violet light that curdled all colour. The grass greyed, all the dandelions dimmed and bleached, the scarlet lining of his coat rotted to mildew green, his hands grew pallid, washed old and deathly, and he felt the cold steal up his legs and swallow him. And then as he peered up the eclipse receded and the sickened sun blinked again in a sky of pure brilliant blue.

The Drawing

March 1975

BAZZA LEANED INTO the alcove to pick and peel the cellophane from the glossy black packet of JPS, which he then quickly stuffed into his donkey jacket pocket. He'd nicked them out of his old man's drawer, he was working nights. He'd get a right bollocking or a beating when he got home, but that didn't matter. All that mattered was Susan Davies and Susan Davies smoked.

He'd just skived maths, she had just done French and would be coming out of the door and along this corridor to English. She was in the top set for both, he was in the bottom for everything, which made chatting her up hard work. He'd have the length of the passageway, the two short flights of stairs and half of the next floor before she reached Walsh's room and his plan was to flash her the packet and ask her to meet him for a fag at lunchtime.

The door opened and a torrent of bored kids charged out

radiating in all directions over the squeaky tiles, one stupid bitch dropped her books and was bowled onto her hands and knees by the herd, then the class next door ended and the chaos was doubled. He shoved some weedy mush away, gave him the glare, and then rose onto his toes to peer over the stampede. She was there so he elbowed his way roughly through and up to her side. She clearly noticed but carried on chatting to her friend so he barged against her shoulder and she glanced at him.

'Alright Susan, French alright?' he blurted. She did something with her mouth, opened her lips, kind of smiled and nodded staring straight ahead, clutching her books to her chest. He smelled perfume. He'd only planned his first line.

'Glad I'm not doin' it meeself, don't get the point, there ain't no Frogs round 'ere is there.' She smiled again, a weak, tense simper that spread into a smirk when she flicked her eyes to her mate who was steadfastly refusing to acknowledge him. And now the crush was thinning, they were walking faster, until they squeezed through the swing doors into the packed stairwell. Keeping alongside her, and her bloody crony, was tricky, the steps were narrow, it was worse than the Milton Road end on a Saturday afternoon.

'Wat'ave you got next?'

She huffed and didn't look at him to reply, 'English.'

'I've got TD,' he spouted over her. No one in her group did TD, they did art, she was in the A stream, he was in the Ns. The Ns did woodwork, metalwork and TD and cooking on the other side of the school. He hated school.

When they reached Walsh's classroom her lanky friend muttered something and tottered in but Davies turned and leaned on the wall with her back to the door. She had properly styled brown hair that curled down to her shoulders. She'd coloured it the same as his sister – it had golden streaks running through it that made it glow. Her face was so smooth and completely spot-free, her lips were still stained with a rosy flush from the lipstick she'd wiped off in the bogs that morning, her eyebrows had been plucked into neat crescents that arched over her wide hazel eyes and she had absolutely fucking amazing tits. She had a real cleavage, she'd been the first in their year to get a bra and she always wore her top two buttons open, even in the cold. He gazed at that juicy curve of shadow that began on the pink beneath her neck and then plunged darkening into the folds of her blouse, he could just make out the frilly edge of black lace dipping away. She heaved her bundle of books up and the shadow thickened and her tits bounced and he felt a flush of sweet perfume and hot air.

'Barry—' She paused, shifted her hips and bit her bottom lip. 'What is it you want?'

She was looking right at him so he looked back at her and said, 'D'you want to meet up for a smoke, I've got a full pack of John Player Specials.' He pulled the carton half out, she peeked at it, looked back at him, sighed 'Okay' and then rolled round the door-frame and into Jane Austen.

For the next forty-five minutes he messed about, pretending to listen to boring stuff about 'angles' and 'trigonometry'. All he could really think about was Susan Davies's tits.

It was a bit nippy, they both had their jackets on with their collars turned up and their hands in their pockets. They didn't speak as they marched over the puddled netball courts towards the cavernous maze of bushes, but as they stooped in and threaded their way through the thicket his elbow brushed hers a couple of times and he muttered, 'Sorry.' There were other hushed voices and whiffs of tobacco drifting in the jungle.

She stopped, shivered and pulled her bag round in front of her. He fumbled with the packet and she pulled out a cigarette. He didn't have a lighter, he hadn't thought of that so she dipped into her satchel and pulled out her pencil tin, removed a yellow plastic BIC, handed it to him and they lit their fags. She was holding the box up with the scratched flowery lid facing him and focusing intently on something inside. He thought it might be a mirror. When he exhaled dramatically and held up the cigarette and said 'The best ain't they' she didn't seem to hear him. She just carried on staring into the tin and at whatever was hidden inside. He checked his watch unnecessarily, buttoned and then unbuttoned his jerkin and eventually asked, 'What is it?', trying to sound nice rather than irritated.

She threw him the briefest look but did not reply.

'What 'ave you got in there?' he demanded. He was hurt, she was ignoring him. Things weren't going to plan, he just wanted to ask her out. 'C'mon, what is it? What 'ave you bloody got in there?'

She frowned and twisted the tin towards him slightly but he couldn't see so he shuffled round to her shoulder and looked down her top. Above her massive creamy tits, tucked into the golden lid alongside her lesson timetable was a small square pencil drawing. He craned forward to get a look at it and grabbed hold of the end of the box to keep it steady as she sucked on the fag.

It was a bird, crouched over some eggs. It was glaring at him with black eyes and a hooked beak. Its feet were there but it had no wings or tail showing and it was incredibly detailed, every single feather had been outlined, each tipped with a delicately drawn arrowhead and its eyes lit up with little white spots of light. Beneath it in minute perfect handwriting it said 'A hen Kestrel and her eggs. From Highland Deer Forest by Lea MacNally, between pages 48 and 49.' Then the initials 'C.G.P.'

'Did you draw that?' he demanded without thinking. She exhaled and shook her head, still entranced by it.

'Where did you get it? Who drew it then?'

Sensing his annoyance at last, she snapped the lid closed. 'Someone in my class.' Her voice was cross and snotty.

'Who in your class?' he cut in. 'Who in your—'

'Chris Packham.' She threw down her cigarette, treading it into the mud with her toe.

'That little spaz with the snakes and that?' he exclaimed. And then 'Hey, d'ya want another fag?'

But she was already pacing out into the open. He flicked away his butt and staggered after her. She ignored him and was walking faster, onto the paving and down the steps, 'He's a total spacker.' And then she was gone, the door banged and he was left glaring at his own reflection. He'd got Susan Davies in the bushes and he hadn't even got a proper dekko at her jugs. He'd blown it.

Richards bawled at him as he ducked out of woodwork before the bell but he made it over to the huts that had housed half the fourth year since the roof of the main school started to fall in. He hid in the cloakroom. The lights were out everywhere, it was murky. He guffed, a real eggy stinker, then untied his polished Martens, pulled up his socks and rewound the ochre laces tight around his calves. Class 4GA turned up and bundled into their desks, then rushed out to get their coats and bags, and when they saw him they fell silent. They knew.

And then he arrived with his mummy-cut hair, with his fringe, with his half-mast trousers and shitty Clarks shoes, with his fawn jumper, staring at the ground as ever. He shuffled into the shadows and reached to get his crappy parka and his tatty sports bag. He wouldn't look at him when he grabbed his throat, when he smacked him in the gob, when he punched him again and got his blood on his cuff, and again in the ear. He wouldn't speak when he pulled him down and kneed him in the chest and then kicked him repeatedly in the stomach and stamped on his hands and his pens and when he flung the bag against the door, when he kicked all his books around on the muddy floor and ripped his coat off its hook. He just lay there when he leaned down and jabbed his finger into his cheek, his eye, when he hooked it into his mouth and yanked his face up, when he wiped his bloody spittle into his hair and spat on him and told him to 'never go near fucking Davies ever again'.

The School Teacher

Thursday 18 September 1975

It takes one joule of energy to raise one gram of water by one degree Celsius so what, he wondered, was happening between the dirty white plug and the scaly spout of this ancient vessel, formerly his mother's courtesy of Green Shield Stamps. It was barely audibly ticking, its ancient furred element was being teased by an alternating current of 220 volts, enough to kill a man or start a fire, so why did this kettle take a geological age to boil, huffing without puffing any steam for maybe fifteen minutes?

John Buckley leaned against the Formica top with his hand on the greasy tap poised to pour over the jumble of chilli-cankered crockery, looking at his spectral reflection in the window. It was still dark outside but he could see the corner of the patio, the slabs wetted by overnight rain, and when a weak yellow light went on next door it lit the brickwork opposite and tessellated his waist-up view of himself with rectangles of pale orange and a grid of grey mortar.

He couldn't wash up, the water wouldn't yet be hot enough, he licked his finger and stroked the top plate but the brown crust had already hardened so when he put it in his mouth he could only just taste the pepper. He picked a tiny clot of the gunk from beneath his nail, flicked it into the sink and then wiped his hands on the tea towel.

Now the kettle grumbled, so in anticipation he spooned some tea into the pot and let out a sigh. It was an hour and a half earlier than he normally got up and he felt it; his arms were heavy, he was listless, maybe he was going down with something – he didn't normally struggle to get going. Unlike the bloody kettle.

He nudged the limp aerial upright and then straightened the radio itself, its plastic base grating on soil that had washed onto the windowsill from a saucer shrouded by a crispy wig of long-dead leaves. The dusty tiles had their own expired ecosystem of fossilised flies, dried woodlice and dismembered spiders, which he studied whilst his fingers toyed with the transistor knob. But he couldn't really turn it on with his housemates still sleeping.

There were three of them and him and another bloke taught sciences at the local comprehensive about three miles away. He did biology, big classes, mixed abilities, big mix of attitudes and today was his worst day; all but one of his lessons were with the bottom-set pupils, most of whom were beyond hope biologically, certainly in terms of passing their CSEs.

They weren't all troublesome, there were some nice kids, polite, quiet, they did their homework, smothered it with felt-tip drawings of cells. But as he'd told his parents at Christmas 'they had a nucleus but their genes weren't genius, they had brains but just not membranes and osmosis simply wouldn't go in, they had Choppers rather than carbon cycles and flares rather than nucleotide pairs. And some were plain prokaryotic.'

They'd all failed their local grammar school entrance exam and been streamed, deemed pretty much non-academic; their lot was basic education and as most of them were working class they were not scheduled for scholastic progress.

The best pupils in the top stream did get entered for O levels, an increasing number went to college and a few got to university. It was tough, there weren't even enough books to go round. It had to change and he wearily accepted that would be largely down to his initiatives and those of his cohort of ambitious young colleagues. But in only his third year as a proper teacher he was still inexperienced, still making mistakes, learning on the job and unfortunately getting a few of his lessons from unruly yobs.

He was also twenty-six, skint, didn't have a girlfriend and didn't drive. He normally got a lift in but the earlier start meant this morning he was on his bike, or rather his dad's bike. 'The black beauty' his father called it, but whilst it might have been a fancied and admired thoroughbred in its youth the withered machine was now a recalcitrant hag that had lost a grip of its gears and chain. It was mechanically incontinent and there wasn't a day that didn't begin in the staff toilet soiling the basin with a slurry of spent Swarfega and his trusty nailbrush.

He dropped the spoon in the pot, switched the possibly simmering kettle off at the mains, flicked the light and went out, carefully closing the door quietly behind him. It was lighter than he had thought, a chemically clear sky sparsely dotted with clouds slipping down to form a thick charcoal scum on the western horizon. He relished the dry smell of city rainwash and nudged his bulging bag of exercise books onto his back, buttoned up his denim smock and pressed his woolly skullcap onto his thinning crown, then tucked his hair behind his ears, tugged at his beard and pushed the ominously ticking beauty out of the gateless gap and off the pavement.

The flabby chain had disembowelled the machine twice before he freewheeled down the hill behind a bus and peeled off up the tarmac ramp to the scrap of wasteland where he was meant to have been ten minutes ago. The kid from his G stream class was already there, stood out in the middle of the field. He pulled off his cycle clips, stabled the bike behind the garages and walked out a short way. The boy raised his arm and held it for a second before lowering it and turning away. What did it mean? It was more of a salute than a wave. Did it mean 'Hello', 'Wait a minute' or 'Stay there'? He remained where he was, his feet already soaked by the rank turf, his hands blackened with chain oil. He could smell dog shit.

The boy was bright but struggled in the classroom and he was impossible to connect with other than through the things that clearly obsessed him. Shortly after they'd first met he'd got very excited talking about his egg collection. Later he'd brought in his fastidiously detailed diaries, full of maps of all the nests he'd found, diagrams of insect flight patterns and tables noting the lengths of all the grass snakes he'd captured. One of their shed skins had fallen from the pages and he'd got upset when it crumbled on the desk. It was 'evidence', he'd said. Then he'd produced a huge bag of owl pellets and they'd begun to analyse them once a month, breaking them open and counting the skulls of the birds' prey. The boy's mum provided the donuts and made him tea in the house that her son had turned into a museum to hold all the things that had died in his zoo.

And then on Monday he'd hung around until the class had emptied and told him about his pet Kestrel. He'd stood fidgeting in front of his desk and whispered an invitation to come and see it fly whilst lining up all the books, pens, chalks and the blackboard rubber. It was a rehearsed speech and as soon as he had stuttered through it he dashed out, fled.

Christopher was kneeling and then walking away. After about thirty yards he stopped, fiddled about, and then turned and blew a whistle, which rang out sharply to start the game. Simultaneously his bird took off and he could hear the tinkle of its bells as it flitted low over the sward and swept smartly onto his outstretched arm. He walked it back and repeated the procedure twice more. The teacher picked his way forward and got his binoculars out, smearing his smock, bag and Karen Harris's green homework book with oil in the process.

The next time the boy threw a lure out onto the ground and snatched it away the instant before the Kestrel arrived, causing it to shoot up and begin to circle around him. The falcon idled for a while but then sped up and began to swoop in more nimbly, its bells falling silent until it whipped up in the air again and glided around before rolling over for another pass.

He edged cautiously closer so that when the bird was at its furthest from the boy it was floating over his head. He didn't need his binoculars now, he could see everything so clearly that he was soon totally absorbed in this incongruous cameo – the pair of young innocents playing together, sweeping through the motions of an ancient art with obvious joy and precision, an incomparable union between their two species that was wholly uplifting, rare and lovely; the free-flying wild thing tethered only by fragile trust to its keeper, but all playing out here on such a tatty stage. On this spoiled green patch squashed between the grisly squalor of the estates they spun, swooped and glowed so very brightly.

There was a lick of wind and the bird broke off and slid out over the slope, down towards the woods, circling in front of the pale brick flats at the top of Kingsdown Way near where the history teacher he fancied lived with her two kids. In response the boy's whistling immediately intensified and he whirled his lure vigorously, worrying the air with sizzling arcs. Then he began to jog down the other side of the hill towards the errant Kestrel but he hadn't got far before it headed back to him, looping in gently with flaps and glides until it was over his head where it stopped, stood in the air and hovered.

The Bird

Thursday 18 September 1975

HE SHOOK HIS wings and then drifted, lightly pirouetting to reveal their dark pointed tips, his tail trailing, a few strong beats, then saucering sideways, a flurry of flaps and then he made a dipping slide up into a stall. There was a split second of stillness before he unfurled into a symmetrical T, his tail fanning full, head down applauding the air, with rippling edges, standing up and aligning, flexing, and I could see it all, every spot and bar, all his rosy creams and soft reds, his neatly folded feet and shining black eyes, dangling on a line tugging against gravity. And then he relaxed his feather-hold, closed the motion, dropped lower, bounced and balanced on the weight of air, fluttering in his element and as he hung there on a heavenly thread just for me, all the purpose and all the energies of our lives fused to produce a moment so perfect that everything tickled in my chest and I felt my heart rocking in its cradle.

He plummeted down to the lure with a vicious thump and mantled it under a jacket of his rowing wings as he tried to drag it off. And he screamed, quivering in the dew, his thorny wet feathers all erected, looking like a crazed bonnet, tail spread, beak gaping, wild. When I crawled up he collapsed onto his chest and refused to step onto my baited glove. I had to unpick his toes from the battered tassel that decorated the lure whilst he bloodied the back of my hand with sharp pecks – he was far too hungry. The old Salter scales had read five and three-quarter ounces and he normally flew on six and a quarter.

By the time I'd prised him off I was completely soaked and he looked completely bedraggled – and when I got on my feet he lunged into a terrible bate. Mr Buckley had come over and he was just hanging there fanning the air, flailing about. I tried flicking him up but he let out a breathless cry, flapped himself to a standstill and then lay across my palm all wrung out and fiery-eyed. When he scrambled upright he stamped on my thumb spitefully and pinched me through the leather with his talons. And just as he settled Mr Buckley reached out to try to stroke him and he disintegrated into a jam-jarred moth of smashing wings all over again. When I eventually got him back together a dog appeared and he froze bolt upright and tight-feathered, jerking his head to keep his bulging eyes on the circling mongrel.

He had never looked so utterly wretched. Even so Mr Buckley seemed to like him and told me a bit about an owl he had kept in Norfolk. But then he asked me where I'd got him. I tried to lie but stammered and blushed like mad so I know he didn't believe that I had found him on the Itchen towpath, but luckily he didn't want to know anything more. And I was glad that he hadn't stroked him because his hands were smothered in grease. So was my biology exercise book when he tossed it onto my desk later that morning. I'd only got a B+ but, more importantly, by using my black Tempo I managed to draw some blotches to mirror the oily finger smudges and restore symmetry to the cover. I was pleased I'd trusted Mr Buckley.

I legged it home at lunchtime, sprinted up the driveway, kung-fued the shed door out of the way and leapt up onto the lawn where I paused, gasping, with my hands on my knees spitting out stringy phlegm. Then I swallowed my panting and peeped into his aviary. He was on his block, bobbing his head, the psychotic villain had become a naughty rascal, he had recovered his composure. When I went in he tipped his head upside down in a playful manner and then bounced down onto his circle of sand and struck at it with his foot, snipped into his fist with his beak, which he shook in disgust, and then hopped back onto his block where he sneezed. I felt tiny specks of his snot hit my cheek and was hugely relieved that he had recovered from his horrible spat.

Last night I'd picked up a dead robin from the gutter in Woodmill Lane and although it must have been clipped by a car it was unmarked, there was no blood and every feather had fallen perfectly in place as I'd stroked and teased it in my palm. I fetched it from my bedroom, snipped off its wings with my mum's dressmaking scissors, then flicked some loose down from its bare flanks and watched it float down onto the mahogany vinyl on the kitchen floor.

I collected wings, dried them by pinning them out on a cork tile, one open, one closed, and then made special card envelopes onto which I wrote all the data – species, scientific name – which I underlined, the date, time and place of finding and any other notes I felt were appropriate. I had eighteen envelopes plus a shoebox with some bigger specimens in. The goldfinch and greenfinch were glamorous but the outstretched swift's wing that I had cut from a corpse collected from the pavement at the foot of Gaters Hill was my favourite, like a feathery knife that whooshed when I made it slice the air. I fantasised about finding a dead kingfisher and getting its wings; they would be my prize exhibit. I'd even considered sneaking my air rifle down to the stream behind the pitch and putt where they nested but they rarely settled long enough to put a bead on.

I'd also tried dissecting a magpie's wing like those I'd studied in a cabinet at the Natural History Museum. The smaller covert feathers and skin had been completely removed so that only the alula, the primaries and secondaries were left attached to the naked bones and all the anatomy was neatly labelled. Mr Buckley had loaned me a scalpel and I'd spent ages trimming away the flesh from the wing, which slid around on one of the best dinner plates and then off onto the draining board. I had the cold tap running to wash away the scraps and had filled the sink with a shimmering scum of blue and green feathers, but in the end the frayed remains looked more like they'd been chewed than scientifically prepared.

I was really disappointed with my efforts but I'd still stretched and pinned the wreckage out. Then flies had got onto it and when I sprayed their eggs it went mouldy so I'd chucked it out of my bedroom window. Later I'd spotted Phillips's cat scuttling through the fence with the manky fan in its mouth.

The wingless robin filled my fist with just its tail tip poking out beneath my little finger. I held it towards him, he bobbed his head, looked once and then fluffed up. I pulled all of the tail through, adjusted my grip to keep the head and body invisible and offered it again. This time he took slightly more interest, tightening his cloak, waddling on his block and ducking a couple of times whilst frowning at the protruding tuft.

I brushed it across his toes but only when I slowly revealed the whole thing did he finally snatch it up, bate away and hunch over it whinnying, trembling and kneading the corpse. He shaved off the feathers and tossed them away, each tear making a quick dry sound like a soft rag being ripped. One or two of the fluffier ones stuck to some glassy blobs of flesh enamelling his beak and he had to flick his head repeatedly to free them and when I flopped down I could see him drawing at its chest, gulping rough chunks of ripe maroon, tiny bloodsparks flecking the sand.

A bone-snipping tug severed an entire leg, which he swallowed after a ludicrous bout of neck-stretching and winding, the foot finally disappearing into his dry mouth as he blinked hard and straightened his whole body with a shiver. Then he returned to the head, which he crunched and devoured far more easily.

After five minutes nothing remained but a fleece of flouncing down and a mangled patch of sand. He jumped back to his perch to pick and pluck at the pink stains on his toes and when he squinted and began to lower his head I offered him the edge of my folded finger, which he willingly stropped his beak on. I rolled the paste of deposited meat into a bogey-sized ball, presented it on the tip of my pinky and he deftly nipped it off. I was already ten minutes late for double chemistry. So what, I had all the chemistry I needed right here. We closed our eyes and slept all afternoon.

The Lessons

May 1973

In the aftermath of the song-bursts the hush was so much bigger. Then the thrush's silver-throated voice fell like pocketfuls of marbles down a church staircase, a richness of sharp sparkling sound, all brilliance spilling bright, shiny and gone quick and made by those silences between, which were more important than all the pealing tings and musical chimes sewn together with harsh scrapes and little squelches that fire worked from the treetop, new-leaved and fizzy. The song smelled of hailstones, of cold tea with no sugar, and whilst the short pauses, the random repeats, the flippancy of its disorder, the variety, were too great and too fleeting for him to memorise it held him there in the shabby thicket, wrapped out of sight in the blanket of browns, out of bounds and late.

The high wooded slopes of his school grounds were sloshing with a milky foam of ramsons, cresting and breaking over the muddle of trunks and washing down through wire and iron jumble to a mauve pool of bluebells, which drained out trampled into the stagnant end of the tarmac playground. A great deadness of bracken spread bleached and corroded behind him and through it the pallid bars of hazel poles rose into a thick net of their whippy twigs flecked with bright beads of gleaming green, their spiked buds beginning to spill the most vivid palette of the woodland's year.

They'd all been belled back inside, the smokers the last to saunter in, leaving just hidden him and the lone thrush still shouting over the divide. As he listened he knew those beautiful notes would hit the classroom windows and fall unheard on the paving below, to be trodden into crisp packets and the sticky stuff from Wagon Wheels, the chocolate chipped from Curly Wurlys, and get blown into the concrete corner drain with that gutted tennis ball that lived there. Birdsong was not a lesson to be learned in that cauldron so he resolved to study it in the safety of solitude. He'd tease apart its subtleties and nuances, make some notes in his rough book and copy them up neatly that evening. Then he'd wait for the lunch bell, sneak home and return that afternoon for double art.

He folded a leaf and the slippery sap cloyed on his fingers, he licked it – bitter – and threw it down to the patio of shining greenery, the flattened fans that spread all around his scuffed Clarkey shoes and illustrated his fresh history of fidgeting. Too sensible, his nylon socks, granddadish with olive lines and tan diamonds, the shinflash of flesh below dead-leg trousers polished at the knees, with empty pockets, stuffed with a shirt and a vest and pulled over with a jumper from last year's sales, stretched sleeves and loops hanging from the cuffs.

The bird had stopped its broadcast and in the distance he could hear the muffled sob of hymns from the hall and the sports teacher's theatrical snarls from the field, goading, rude and unfair, railing at the incapable, feeding the raw and cruel persecution of the few competents and their gangs of popular friends all afforded immunity from ignominy due to an ability or interest in kicking, hitting or catching a ball.

P.E. and sport were just hell. Get mocked because your bag wasn't Adidas or Puma or Gola, or wasn't the right size or shape, get ridiculed because your kit was old or your boots weren't Franz Beckenbauer with three dayglo green stripes, get jeered at for slipping, stumbling or falling, get a finger screwed into a temple for not knowing a rule or called a monger for fumbling a catch or a pass or be labelled a 'homo' for accidentally brushing against someone in the showers.

Then get whipped by rat's tails, towels knotted and soaked in water, or have everyone shift their stuff to other benches because they spot your eczema and wince and scream and point and then tell all the girls who already think you're a retard. It was better to forget your kit and spend the time picking stones out of the newly sown pitches that they had spread over the old brickfields where he had hunted for hedgehogs and slow-worms and seen a deer one winter's morning whilst on his way to the dentist's. It had massive ears that swivelled forward and back again independently before it bounded daintily into the soft hummocks of brambles to play a far more beautiful game.

He saw a rat, briefly. It slunk down the underside of a log, vanished and then bolted across the path into the ivy, a rich brown blur, a creature without real form, a sum of scant parts that combined to give it an identity. He wondered how he knew it was a rat, a brown rat in fact; he'd seen no tail, or nose, ears, face, just a shape shifting fast with a sense of colour. In his mind he could conjure the picture of the rat in his *Observer's Book of Wild Animals*, it was a photograph that had been coloured in, the rodent faced right, had prominent white whiskers and a thick tail whose tip was hidden behind a leaf.

He could always recall pictures, see them in his head, he could see the page number too and its name, Brown Rat and beneath, *Rattus norvegicus*. But it didn't look like the creature he'd just glimpsed at all. He liked the way you could 'sense' animals, feel what they were, even what they were going to do next. He always knew when a bird was going to take off, he couldn't explain how, it was as if he could see it happening before it sprang into the air.

Sometimes he could tell the difference between apparently identical individuals of the same species too. Once he had put a whole line of lackey moth caterpillars he'd coaxed off the bark of the pear tree in the garden into a flowerpot. But they had scaled the sides and had begun to circle the rim, their narrow stripy bodies inching forward in a line, all following each other, processioning nose to tail, their blue grey heads marked with two happy oval eyes like those on his 'smiley' badge.

He'd been watching them, teasing them onwards with a stalk of grass like a ringmaster whenever they slackened their pace, when his mother appeared. She knelt down, studied the fluffy circle through her pointy glasses and then began saying that they were 'like little robots, all exactly the same'. He'd told her she was wrong, that they weren't the same at all and she'd scoffed and asked him to pick a caterpillar, look away for a minute and then turn back and point out the same colourful circling larva. He did, easily, three times. It was a sinch, that particular caterpillar was pausing more frequently and raising its wobbly head a bit higher than all the others, something he'd noticed as soon as their funny carnival had begun.

The rat had scuttled into a dump of rusty old paint tins and broken plaster, a green-stained hummock bejewelled with smashed glass and slippery shards of tiles. He kicked the crusty tubs, hoping to flush it out, but nothing stirred. Using his toe he cautiously teased the edge of a sheet of corrugated iron, which suddenly catapulted up when some strings of ivy broke free and, when he wrestled it over, the dry bed beneath revealed a motley collection of small slugs, worms, a centipede and an instantly agitated devil's coach horse.

It 'scorpioned' at him as soon as he poked it, flexing its jaws and arching its jet-black tail over its writhing back, waving its beaded antennae before attempting to flee like a giant earwig into some chunks of dry soil. But he cornered it and gently pinned its head down with his fingers before picking it up.

He felt its jaws pinching his thumb as he noticed two small bright white waxy tufts waving from the tip of its squirming abdomen, which was simultaneously leaking fluid that slipped over his nail and down the insides of his fingers.

Then he smelled it, a powerful, acrid, spiteful and clinging odour, far sharper than the nip emitted by ladybirds and the warmer tangy scent he got on his hands whenever he found a paralysed ground beetle spinning belly-up in his driveway and tried to rescue it. This truly stank but rather than throw it down he brought it up to his nose and breathed slowly in, smarting and saturating his senses with the noxious perfume, simply because he didn't want to forget it. It was a pity he couldn't do an O level in beetle smell. Or rat identification, or birdsong.

He washed his hands after lunch with the mucky cracked stone of soap that sulked turtle-like beside the enormous sink in the art room, but he could still smell it and couldn't resist occasionally testing its persistence by surreptitiously sniffing his fingers all afternoon.

He drew and painted a black sun with orange clouds raking across it and in a space age script of his own invention wrote 'The End' symmetrically beneath it. When Mr Tucker finally stopped revelling in the giggling bouquet of girls draped around his desk he ambled past and merely nodded before embarking on a verbose and flattering appraisal of Deborah Barr's felt-tip monstrosity of a vase of flowers she'd brought in especially.

January 2004

'I just wasn't able to be that emotionally energetic or get that excited, I couldn't be bothered to pretend to be enthusiastic just to get their attention. Not about something I had absolutely no enthusiasm for, or absolutely no interest in.'

Before she could speak he went on, 'And, well, it's not like it's about being like that once, to one person, is it? It's about being Mr Bloody Bubbly all the time, to everyone, because it's not a singular or linear thing ... it's like, I don't know ... it feeds on itself, it expands, reciprocates, grows into a monster of happy fun which stamps around people like me having the time of its life, until it gets bored and then kicks and beats us just to reassure itself that it's feeling so brilliant.'

He was much brighter today, more confident, he was speaking more quickly and spontaneously. He seemed almost carefree compared to his usually critically considered self. She wondered whether this might reduce or alleviate his caution, whether he might be pushed a little further than normal. She asked, 'So you hid from the happy fun monster ... ?'

He chuckled. 'No. You can't can you, unless you're a hermit or something. Every day we have to go out and dance with the monster, mix in, have a chat, socialise, work, but it's tiring and it gets you down in the end, you have to be able to get back to yourself to rest up and reload for the next assault. I realise now that's what I used to do, go to physical places, places to be alone in my world, that was part of what all that time in the woods, the fields, by the river, in my bedroom was all about. I thought I was running away from the domestic anarchy but it was more than that.'

Again without prompting he went on, 'But let's be honest, I don't think that's much different for loads of people, everyone needs a break don't they? Don't people do all that exercise stuff, yoga, I don't know bloody what, just to get some headspace? What's the difference?'

He was definitely in a better frame of mind so she began to nudge him. 'But that isolation, that need to find safety in solitude must have impacted on your ability to share friendships?'

It worked; he instantly sat up and squared himself into the chair and after his first hiatus of the morning replied to his shoes. 'Yeah, of course it did, it has. I've had very few proper friends. But it's better that way, I don't need to know lots of people socially, to know them well enough to be comfortable enough to be able to just meet up with them, for no reason, just for the sake of seeing them. I'd rather give more to and get more back from one or two people. I suppose that makes me antisocial but I have lots of people who I can talk to, that I enjoy talking to, but in a very partitioned way. I'll talk to them, only sort of know them, in regard to a single subject, birds, art, music, football ... work, whatever, but I wouldn't talk to them about anything outside that compartment. Or about my life, or about what I think. That wouldn't be appropriate at all. I like them in that one regard but I don't need them outside or on top of that.'

He drew his feet up tight to the chair legs and angled them straight. 'Maybe that sounds bad, but it's not, it's not like that at all. It isn't. I very consciously, constantly, try to offer them something back in return for their company, and ultimately I don't have a choice about any of it anyway.'

If not wholly constructed his ability to empathise had been learned, she knew that, but momentarily she felt a pang of sympathy. It must have been hard to start developing that strategy and probably next to impossible to maintain it with any long-term functional efficacy. 'What about your partners?' she asked. 'How have you managed those relationships, given your need to actively adapt to try to make yourself more acceptable to them?'

There was a longer wait before he began, in a lower, lesser tone, a modified mood and then he said, 'Honestly? ... in the short term fine, but over time, clearly not so well. Not well at all. My fault though. I've been too scared to be completely open with any of them about what's been going on underneath everything and that has meant there's always been a burgeoning graveyard of skeletons rattling around which I've needed to constantly try to corral. And that has been ... um, terminally exhausting and I've eventually failed. But what's ironic is that I only seemed to mess up when I thought things were going really well. Things have only fallen apart when everything seemed secure. Well, seemed secure to me. Which doesn't mean much does it ...

'For me when anything gets too good is when it inevitably goes bad. Whenever I've got scarily anywhere near normality that's when it's gone wrong. Maybe it's because that's when I've let my guard down, I've got lazy, forgotten what I am, forgotten that I'm vulnerable. Made mistakes. Like trusting people, like having expectations of their behaviour, their commitment. Humans can't be trusted with expectations. Or completely trusted full stop. Animals can, not people. You can lend them a fiver but you can't give them your heart. And that's why I'm here isn't it?'

He folded his arms and cleared his throat.

'I'm here ...', he sighed, 'because the thing that truly loved me died. Because I then collapsed and forgot that a normal human–human relationship is too dangerous. Because I was distraught, desperate, I trusted someone. I loved them. I began to think that I was going to be okay inside something that I could never actually ever have and now I'm paying the price for that. No, now I've paid the price for that. Actually no ... I nearly paid the price for that.'

6

Anna, My Other Angel

The Paper Round

January 1974

THE STREETLIGHTS HAD sparked up prematurely, purring apricot in a foul-tempered sky, and joining their dots made pleasing curves up Avon Road. The familiar cadence of the classified results had been audible through the letterboxes of the last two bungalows where he'd posted newspapers and men sat gnawing pencils and praying for score-draws and then ripping up coupons and lighting a fag. His dad 'did the football'; every Thursday night someone knocked on the door halfway through Top of the Pops and he'd get up and hand over a slip of squared paper marked with random biro crosses along with a few pence. Mysterious people apparently won millions, proven by the ecstatic faces of bobble-hatted pensioners from Lichfield and second-hand car salesmen from Halifax who stared out from the same papers that occasionally revelled in lurid stories of pools winners' shambolic spending sprees and celebrated their predictably miserable return to poverty.

It was clearly an idiotic waste of money; his dad had absolutely no interest in football, didn't even know who was in the First Division, let alone who was any good, he just spaced his Xs haphazardly down the grid and made Littlewoods richer. Why didn't he just stop it, put the money in a tin each week and then do something with it? Why were they always on about having not enough money when they threw it away on 'doing the football' or endlessly smoking cigarettes?

It had poured down and the spines of the newspapers were grey and crumbling so he folded them into a roll and rammed them through the rectangular jaws of over-sprung, finger-pinching chrome and brass and aluminium mouths, into the carpeted throats of other people's lives, feeding them morsels of facile information and the fabric to flatten in the trays of budgie cages or to screw up and set smoking as they strained their lungs to blow into reluctant grates piled with the last hissing wet coals whilst kneeling in candlelight on scorched nylon rugs pulled over lino and groaning, before moaning, about the 'bloody Three Day Week'.

He'd got as far as the gate when the door burst open behind him and some old codger bawled 'Oi!' He stopped and they stood glaring at one another; a waft of beer and a cat squeezed out between the lardy legs of the man, who looked down and spat 'Bugger' before stumbling backwards into darkness and slamming the door, the knocker rapping, the streetlight quivering on the dark window glass.

When he dragged the gate half closed flakes of green paint stuck to his wet hands and he wiped them across his soaked trousers before shoving his bike on up the hill. In the next house he found a crumpled note marked with the faintest trace of spidery biro drooping from the letterbox. But as he leaned around his shadow to read it the weak glow went out along with all the other squares, triangles, circles and spots of curtain-coloured lights in the road and the dirty silhouette of the city's skyline lost all its sparkles. 'No paper today please' is what it read, and on the back 'No milk today please'. As he stepped off the doorstep an old woman's face swept away like a ghoul as she drew back from the window, a grey oval hovering for a moment in the dark but then vanishing as she sank into her tomb.

He finished the round, pushed his bike away from the newsagent, across the pub car park, stopped by the electricity box fence and pulled a mushy Toffee Crisp out of his clammy pocket. Once, years ago, he had told his gran that they were his favourite and from then on every Saturday she would hand him one with a self-satisfied smile. This one he had forgotten to add to the biscuit tin full under his bed. He picked off the soggy wrapper and flicked it away, took an apprehensive bite and chewed on the paper he had missed in the dark. Then the streetlights came on and he saw something rolled up inside the fence.

The Pin-Up

January 1974

ANNA. IN A mirror, double Anna. Miss Anna. The tip of my tongue pressed on my teeth, her hair, her hands, her mouth open, her reflection gazing, big brown-eyed so fast into my blazing mind, my whole head boiling, dizzy. She held her wrists across the top of her breasts, rounded, perfect, symmetrical and with nipples. Her blouse was open, her belly button and there between her legs Anna had a cluster of hair, wispy at the edges but woven into a thicket, brown and disappearing deep into the shadow of her thighs. Hooped stockings, red, white and black. Anna, bloody, bloody hell, Anna Bergman, Jesus, Anna.

A car trundled round the corner so I stuffed her into my sodden parka. Rain pinged on the corrugated iron fence where I'd found her, the old wreck stopped, then its indicator blinked, come on, come on, another car passed, I heard the handbrake click, the engine over-rev and then slowly, so slowly it edged out and up the hill. I pulled her out and skimmed quickly past Shelley, Shelley lying in the window, Shelley drinking, Shelley pulling off her pants, Shelley with her legs open, dappled Shelley with her bushy triangle of hair, past cartoons, my wet hands fumbling the pages, Michael Winner, page forty three, come on, forty thr-e-e-e ... Anna, my Anna, on her back her legs opened wide, stockinged, draped over fringed cushions, her feet on the arms of a fancy chair, a necklace between her breasts. Anna with her arms spread, her hands against a window, looking worried, sad, her naked backside cut with a thick black shadow where that hair was and then Anna ... sat touching her fringe, her right hand, her face looking down, her skirt, checked skirt, pulled up, her breasts, a ring, a bracelet, red, black, white, her nails, shining. Anna leaning back, her nipples bigger, curly blonde hair, Anna kneeling with no breasts and no hair and then, then Men Only Girl: Miss Anna Bergman, beautiful, two pages, looking, big eyes sparkling beneath her golden hair, holding her necklace, half kneeling with her blouse wide open, with her breasts, with her nipples, beautiful with that nest between her legs, with her white socks. Anna.

I spent nearly two months with her, I saw her every day, sometimes more than once. I met her friends, Sidsel, Shelley, Nikki, Britt and Divina. They were all naked, all exciting, they all had breasts and hair between their legs, they all looked at me as if they wanted something, or were scared or like they knew me.

All except Britt who looked stern, cross and disinterested. She wasn't as pretty either, her breasts were shaped differently, her nipples darker and larger, her hips really bony. I hardly ever looked at Britt.

Divina was black with a huge frizzy Afro and tightly curled, thick shiny hair between her legs. She was kneeling on a fur-covered bed with her blouse pulled up. All their clothes were pulled up or open and they all had loads of necklaces and they all lived in rich houses or mansions with posh chairs and beds and none of them looked anything like any of the girls at school.

But it was Anna, the eight photos of Anna that I liked best. She was special, she was the most beautiful woman I'd ever seen and somewhere she actually existed, she was real somewhere, she walked down roads, spoke, slept, ate things, lived.

Sometimes I'd meet her quickly on the way home from school, waiting for all the other kids to buzz off first, and then I'd cycle back to see her again after tea, in the dark, with a torch. I'd take her through the fence and hunker down behind the electricity station.

Sometimes people would walk past towards the pub, I'd hear them over the hum of the grey boxes and see their misty conversations rising into the pinch of the night and as the cars turned in the car park their headlights made Anna glow, swept her tanned body with gorgeous light, filled her secret shadows with flashes of mystery and I'd learn the photos, say them, whisper them into my memory so I could see Anna at breakfast, in assembly, during chemistry, in the back of the Maxi and out in the woods and fields. And in between loving Anna I read the text and learned a new language. And then one day at the end of February she disappeared.

The cold scratchy whistle echoed loudly and through my binoculars I could see the silhouetted beak sputtering just out of sync with the bold trio of notes that rang out monotonously and raked away into the glassy evening. I could also faintly hear the jazzy jingle of Nationwide starting in the house alongside the big birch tree where the mistle thrush was performing. It had been there for a few weeks, Anna and I had been distracted by its excited rattles and scolding pursuits of the jackdaws going ape around the chimneys – it clearly had a nest.

My *Observer's Book* said I'd find it 'in the fork or on the branch of a tall tree' so, having checked that Anna was still waiting for me in the corrugated iron fence next to the substation, I leaned back and began scanning the row of trees for the bushy tangle. Indeed, the trusty brown bible described the structure as 'rather bulky and conspicuous, built of twigs, roots, grass, moss and wool' and said that the eggs were 'sometimes laid as early as February'.

My breath kept steaming up my binos and I could find nothing in the steely grey crotches of the birches as I searched methodically through their wafting mist of purplish twigs, carefully checking every crook from right to left towards the still-bugling bird. There was no sign of a nest but the tree it was perched in was stockinged in a thick coat of ivy; before it got dark it was worth a climb.

I opened and closed the gate latch silently with both hands and snuck down the cracked concrete sideway, ducking under the kitchen window, before climbing over a plastic dustbin onto the gritty felt roof of the shed and into the birch. The thrush began rattling somewhere above me as I tugged my way through the necklaces of dusty creeper, my fingers stretching to reach the branches, my toes squeezed into the crutches and cricks where they met the trunk. It was an easy climb and I soon reached the flimsy crown where the ranting bird had been joined by its agitated mate. Both began chattering angrily as they dived at me and pinballed from perch to perch.

The nest was there, in the shadow of the ivy, moulded into the loins of the main trunk, and after one more thrust I was directly beneath it. I panted, spat out a twig, rubbed my eyes, adjusted my footing and then reached over my head and into the cup. It was deep and contained three warm eggs; she must have been on it all along. One by one I removed them and turned them gently with my thumb to examine their patterning. Each was pale cream, tinged with a wash of tan and specked with spots and blotches of dark reddish brown and mauvish grey, slightly bigger and rounder than the song thrush I'd collected from the school grounds.

I put the third one back in and ignoring the swaying tree and the increasingly desperate birds relocated the second egg, which I'd judged as the most attractive, and popped it into my mouth. This was how I carried my booty safely down, not trusting my pockets and unable to single handedly place them into the plastic egg box I'd hidden in the fence with Anna.

Once across the road I brushed myself down, shook the debris from the hood of my coat and down my sweaty sleeves, rubbed my hands clean and bent over to slip the egg out of my mouth. I wiped away a mess of dribble and examined the glistening prize. It was lovely and should definitely 'blow', as the clutch wasn't yet complete and it would still be yolky and not contain the partially developed embryo of the young bird.

Kneeling down I reached into the cubbyhole, took out the tub and nestled the egg into a cosy hole in the cotton wool. Then I pulled my binos out and focused on the tree. In the dark the thrush grew bigger and was back on its perch, not singing but preening – they hadn't 'deserted'. Then I squeezed the new prize into my pocket, checked there was no one around and reached in to Anna. But she wasn't there.

I stretched deeper into the rusty hideaway, my fingers tapping around for the magazine ... nothing. Maybe I'd pushed it out of reach with the egg box? It was too dark to see so I walked home and immediately ran back with my dad's rubber torch. The chamber was empty. It was unbelievable but whilst I'd had my thieving hands in that nest someone had nicked her. Anna was gone and I never saw her again.

I checked the cache the next day after school and several more times throughout the spring in desperation but in my heart I knew it was over. And for years afterwards of all the faces that appeared to me out of the past, the one I saw most clearly was that of the girl I never ceased to dream about, she was the first earthly love of my life and I never forgot her name.

The School Trip

November 1974

ZIGZAGGING RIVULETS FLINCHED down the window, jerked back by their twinkling volume and the queasy progress of the coach, stampeding between traffic lights on spongy tyres that hissed on the glittering streets. She tried to trace their course with her finger, but they fled fast down random routes leaving constellations of vibrating drops scudding into their tracks where they dashed into wobbling lines of amber and red, thickening and thinning with the passing of the long line of overtaking cars.

She found parts of the window where raindrops never seemed to settle and pressed on them with her fingers, watching her small pink nails darken with blood to a deep rose and then grow paler near the tips when she pressed really hard.

When they reached the main road the driving improved, the coach merely lurching across the lanes dodging lorries and queues with loud exchanges of horns. It was soon travelling too fast for the rain to stick to the glass and instead it had begun to mist up on the inside, fogging out the world and isolating her with only her friends and enemies.

On her lap she had placed her small red copy of *Pride and Prejudice*; she was in the top group for English. Beside her Sandra, her best friend, was reading *The Silver Sword*. She was in the next set down, which wasn't fair but Miss Walsh had said that she might get put up the following term so that she could do the O level rather than the CSE. She hoped that would happen; they could sit next to each other and share their notes like they did in maths and science.

Sandra was wearing nail varnish, which wasn't allowed but as it was just a weak wash of petal pink no one had noticed yet. They definitely would spot it though, and then she'd be sent to clean it off. It was all chipped anyway because she chewed her nails. And if she had been pressing her fingertips against the glass she wouldn't have been able to watch her quicks changing colour.

All the girls had their better clothes on, not their party best, but not the things they normally wore to school. There was no uniform; you could buy a tie and a blazer but only the new second-years did. But there were conventions, which would be discussed during every tedious assembly. Mrs Dunlop, the aged deputy head, seemed to have little else to contribute other than when choir practice was scheduled. She would drone on about the length of skirts and the height of heels and 'discretion' and 'decorum' and 'self-respect'. But they didn't do decorum at C&A and no one with any self-respect would be wandering around dressed like her. And her current obsession was with nail varnish, eyeliner and jewellery, all of which she loathed and defined as inventions of Satan devised to ruin the reputation of the school – which had no reputation.

Not that it mattered; she wore long and full skirts that she and her mother made, no make-up – she wasn't keen on it – and a tiny gold cross threaded on a thin chain that had belonged to her grandma.

But Sandra had begun to change since the summer; she had a lime green chunky knit cardigan on, a bright floral blouse with a frilly collar, a candy-pink skirt and thick black tights. It was loud and, combined with her explosion of curly auburn hair, she looked a bit like one of the girls on *Top of the Pops* or those who hung around outside Chelsea Girl on Saturday afternoon smoking and squealing at the men in lumberjack jackets with furry collars and singing along to the Rubettes. Not that it mattered how she looked.

She turned the book on its spine and squeezed all the pages together to make a gold stripe between the neat leatherette covers. Then she selected a single page of the delicate, almost translucent turquoise paper and flexed it between her fingers to see if its fine edge would glimmer under the soggy bulb she had switched on above her head. It wouldn't so she closed it again wearily.

Sandra was craning over her own tatty book, frowning through her bulbous lenses, stroking the ragged pages down and trying to fold the furry corners flat as she pawed through the text, all pencil marked, underlined – contaminated by the long list of reluctant readers whose names were inked into the front in their best handwriting.

The coach was wretched, the chrome was flaking from the rim of the pop-out ashtray in front of her that she gingerly stroked but knew better than to open. There were accompanying cigarette burns in the split vinyl seat arms, melted black scabs ringed with a halo of singed tan and a tear in the grubby fabric that could be peeled to reveal the wooden skeleton of the seat in front. Onto this a previous passenger had carved 'POMPEY KICK TO KILL' in blue biro.

Suddenly there was a commotion. Everyone was leaning into the aisle or standing up, people were whooping and cheering, the teachers were stood up, shouting, telling everyone to sit down and shut up. It was down the front, she couldn't see through the mob what was going on but everyone was creased up, some were clapping, then a round of applause broke out, followed by groans and more hysteria, and gradually things settled down and slowly the message filtered back to them ... Roderick Sutton had chucked up all down the back of Gilmore's seat! It wasn't worth the riot but it warranted a smirk. Mr Gilmore was the year head and had no apparent sense of humour. He always wore a suit and he'd now be spending the whole day at the museum covered with Sutton's chunder.

After the furore she knelt against the back of her seat and chatted to Pamela Martin about her mum for a while. They both had 'mum and dad trouble'. Pamela's dad had left and hers was drinking too much. They got to sort of agreeing they were on their mums' side but kneeling was really uncomfortable and the driver was ramming the brakes on all the time so she sat down before she was told to.

The teachers were being really snappy, the trip had nearly been cancelled because of the bombings; there had been another one this week but it wasn't in London so they had been allowed to go after a long lecture about responsibility and discipline. The whole summer had been about the IRA or the elections, it was really boring. Her dad just sat there hogging the TV, watching the news with another bottle. Then *World in Action*, then *Panorama*, more car bombs, more men shouting, the army running about, and then a blazing row about whether her mother could turn over to watch *Upstairs, Downstairs*.

Sandra was still pawing at her pages so she gazed over to the seats opposite them. Only one was occupied, it was the boy who never looked at anyone and who every once in a while tried too hard to join in. He was reading too, it was a paperback and she could tell from the plain cover it was an adult book, a proper book. She wanted to know what it was. If it was anyone else she could just ask, but she couldn't talk to him, not like that.

Sandra was his practical partner in chemistry and she hated him, she said he was stupid and clumsy, that he couldn't even speak, and didn't understand anything. That he only got Cs and he smelled of mice. But when she had sat next to him herself in biology and sneaked a look through his book it had been so incredibly neat with amazing drawings and no crossings out, and he'd got 99% for his end of term exam.

Mind you, he hadn't spoken a single word to her all afternoon unless she'd asked him something and even then he had merely mumbled to the desktop in reply. At one point she had accidentally touched him and he'd jumped out of his skin like he'd been electrocuted and then he'd blushed for ages. In fact he was always blushing: if anyone surprised him, if he got asked a question by a teacher he would start glowing, his ears became hilariously red and the other boys would howl and flick at them and jeer as he hung his head or skulked away dodging their kung fu kicks and chops.

He was always anxious, like he was constantly embarrassed or angry, and they loved that and never missed an opportunity to ridicule his pitiful shyness. And what was really weird was that she never saw him around, he was either at his desk or sat in his chair in the next classroom, he was never in the corridors, or in the cloakroom, or in the tuck shop queue or walking anywhere. He was never in between, he didn't exist out of lessons, in breaks or at lunchtime. She presumed he was just avoiding them all, it was like he was forced to be in their world when really he only wanted to be in his own.

But then sometimes he'd do something outrageous, completely out of character, like locking their French teacher in the book cupboard, heaving a huge rock of chalk at his geography mistress or once tearing the whole pinboard off the maths room wall. She'd seen him do it, during book break on a Friday afternoon. The other cretins were flicking Blu-tack, pinging elastic bands or tugging at the homework timetables so they hung askew.

He went way beyond that but rather than impress them, he scared them. Maybe he knew that, maybe it was his revenge. Then there was the time he'd terrified everyone with the snakes that escaped from his desk. Mr Buckley the biology teacher had stuck up for him but admitted that whilst the grass snakes were a bit of harmless fun, bringing the adders to school wasn't such a brilliant idea. No one got bitten and he was sent home with them during RE. But for weeks afterwards the boys would screw their index fingers into their temples, push their tongues under their lips and dementedly moan 'snake-mental' at him.

More recently someone had opened his desk and found a dead owl. This time some of the girls had complained theatrically and it was confiscated. Apparently one of the fifth-years found it in the big clanky bins behind the kitchens but Buckley had told her and Sandra that he'd collected it after school and had taken it round to the boy's house and shown him how to stuff it. They'd cut it up and used marbles for its eyes. Buckley liked him but now no one would sit at his desk when they had French or Spanish and Susan Bitch and Caroline Cow and their gang would make a show of walking round him like he was diseased. That's why he sat on his own.

The Fossils

November 1974

Mary in her bonnet, with basket and hammer, upright in a heavy cape with her sleeping dog on the beach at Lyme Regis. Plain, demure and dead at forty-seven, she'd picked amongst the landslips and found the first Ichthyosaur aged just twelve and later the first two Plesiosaur skeletons. But she came from a poor family and as a woman was excluded from and exploited by the rich scientists of the day. She had a fossil shop and sold specimens to museums all over the world and he knew all of this from *Blue Peter*.

He'd met Miss Anning before of course, standing here in the long narrow hall clad to the ceiling with massive mahogany frames, each containing cherished plaques of flattened Plesiosaurs, their beautiful skeletons mounted in a matrix of drab lichen green. It was a fabulous collection but to see it properly he would have needed a ladder. The upper specimens were hung too high and when the sun bled through the skylights he stood squinting through reflections at the rows of ribs and the four long flippers with their leaf-like mosaics of finger bones and their long narrow necks. He was eager to see their skulls, each packed with hundreds of needle-like teeth, perfect for preying on fish it said in his old *Blue Peter Annual*, number eight. He still looked at it occasionally, he liked the story of Tipu's Tiger.

The rest of his class were sat in clusters at the other end, past the marine crocodile and boring Stegosaurus fossil, unsnapping Tupperware and tucking into sandwiches and Smith's crisps and Wall's steak and kidney pies, their collective voice rapidly escalating between the regular bellowing of the teachers, which echoed far more loudly than the quiet they demanded. Idiots.

He'd deliberately left his polythene Co-op carrier on the coach, he didn't want to waste any time in the museum or be taunted because it wasn't the right type of bag. He could eat his squashed sandwiches on the way back but now all he wanted to do was enjoy his favourite Ichthyosaur. It was German, between 187 and 178 million years old and amongst the grate of ribs and backbone were the bodies of six unborn babies.

He pressed his fingers onto the glass; their skulls were outlined in red but were still hard to discern, if he could feel them it would be easier but he could definitely see little snouts lined with teeth. It was brilliant, he loved fossils. He stepped back admiring the 'sea dragon' just as Kevin Banks thrust his wobbling, goggle-eyed head in front of him and announced in a low cheese-and-onion flavoured drawl ... 'It's frothy man!' before sprinting up the hall, skating through the door on the polished wood floor and running, not walking, towards the bogs.

To the left was the actual skull that Mary and her elder brother Joseph had uncovered in 1811. It lay at the bottom of the lowermost cabinet beneath an exceptional and immense Ichthyosaur with the long column of its vertebrae stretched out behind it and looked as if it had just been dumped there, as if it was being temporarily stored. It had a proper label but why, he wondered, didn't the world's first Ichthyosaur skull have its own special cabinet? Maybe the scientists still didn't like the fact that Mary was better at finding fossils than they were.

Squatting down he studied its bird-shaped head, its long toothy beak and its gigantic eye, which he measured with the span of his hand. If only he could see what this marine reptile had seen in the stormy Jurassic seas – porpoising through the surf, pursuing squid and fish, speeding into the depths to escape giant sharks or Plesiosaurs or ferocious crocodiles, dodging ammonites, dying, getting covered in silt, being fossilised, waiting in the dark whilst the rocks rolled and the world rumbled until the land slipped and Mary found you and everyone thought that you were a fossil crocodile and posh palaeontologists studied you and the public talked about you and you were famous. But then someone found T. rex so you were tossed into the bottom of a case to be ignored by schoolkids obsessed with *Planet of the Apes* or David Essex.

That was the one thing wrong with the British Museum of Natural History and it was a whopping great mistake, an unbelievably cretinous botch-up, it was the one thing here that had always, would always disappoint him ... they'd muffed it, they didn't have a Tyrannosaurus. rex skeleton. And if they did, then all those divvies and goobers and spazzmos wouldn't be shaking up fizzy drinks cans, trying to walk like a chimpanzee or run in slow motion. They'd be gobsmacked by the greatest animal in the entire history of the planet.

He crept closer to his classmates and sat down just as Mr Gilroy said to put everything back in their bags because it was time to go to the Science Museum. Boring. And it was, loads of cabinets with broken displays, kids punching brass buttons and yanking levers in sombre, dingy rooms, it was all physics and no biology.

On the journey home the principal source of repeated hilarity centred on the facts that Victor Giles had successfully pitched a penny into the blue whale's spout and Martin Medley had wiped a bogey onto the tip of the Triceratops horn. A prefect had nicked his seat so he had to sit next to Buckle, the white-haired spectacled girl who they had all bullied senseless since the juniors. They didn't share a word of course, the oppressed were even terrified of each other.

So instead he mused over who, if he could have anyone, would be the best mum in the world. Raquel Welch was his favourite and might know a tiny bit about dinosaurs, Virginia McKenna from *Ring of Bright Water* would surely allow him to have a pet otter, although he couldn't imagine her in a fur bikini, the woman from Portswood Pets would definitely let him keep a crocodile ... and a fruit bat, and a green iguana; and Valerie Singleton would probably be cool too, but maybe a bit more strict about when he got back from exploring. And he still liked Valerie from *Land of the Giants* even though she always needed to be rescued by Steve. And the woman from *The Champions*, she was telepathic and had superpowers. By the time they bundled off the coach and scarpered down Dimond Road through a volley of sleet he'd decided on her, Alexandra Bastedo.

Beefburgers, unburned, chips and peas, he thought it really hard with his eyes closed, with as much red sauce as he wanted, he visualised it on a plate, muttered the list again and again into the upturned collar of his coat all along the wall in Cornwall Road but when he kicked open the back door he was smothered by the pong of his culinary nemesis – boiled liver and bacon. His mother had failed to get the message.

The Bird

Monday 29 September 1975

IT SMELLED SOUR, acidy, like some animal's piss. They'd picked the apples yesterday, Bramleys, cookers that tasted bitter and dry when you bit them, when you spat the white puffy flesh out and hurled the fruit hard at the brittle trellis fence, which cracked loudly on impact. All the family had been out in the garden, my dad up the ladder tossing them to my mum who polished them with her apron and passed them to my sister. Now they were all laid out on newspaper across every flat surface in the shed, not touching, drying. They were vivid green with weak blushes of rose where they'd faced the sun, some with fawn bruises and flattened sides. I'd helped load them into the handle-less blue buckets, using my little fingernail to rescue ladybirds that had got jammed head first down into the stalk pit, where the skin had flaked to crispy brown rash and tiny hairy white spots grew, some sort of aphid my granddad had once told me.

They never really dried, they would remain waxy all winter long and next weekend my dad would individually wrap them in half sheets of the *Daily Mail* and *Southern Evening Echo* and store them in the two chests of drawers, the apple drawers. And on Sunday mornings me or my sister would be asked to go out with the saucepan and bring some in and we'd tug at the handles and wince at the vinegary belch and have apple pie again and again and again.

I was in there to check my mousetraps and had caught one of the fifteen or so escapees from my breeding programme. The plan had been to run a mouse factory to provide fresh natural food for my bird, but the inmates had chewed their way to freedom and had been getting into the kitchen and seen darting around the dustbin.

Stretching through the frame of my bike and a ladder I reached blindly for its tail, which, after a struggle, I pinched up to drag the body and the trap clattering through the cobwebby mesh of metal and wood. It was just hanging on by the front half of its head, its eyes burst out of its smashed skull and a lick of blood on the treadle.

I prised up the bar and the soft furry pouch of bones fell silently to the floor landing pale belly up where it glowed blue-grey, the colour of snow in shadow. It was soft, didn't yet have rigor mortis, it wasn't warm but had only been dead an hour or so. I pressed its hard yellow teeth and stroked the blood that spilled from its ripped mouth into its fur, pinkening the grey, blackening the brown of its flank. It gurgled, the tiny chambers of its heart choked with the congealed spume bubbling from its lungs. Its feet were clean and frosted with a down of pearly hairs and I held one of its hind legs and scraped its claws over the back of my hand where they engraved a lattice of dry scratches. Then I slipped it into my pocket and headed for the aviary.

Before going in I paused by the door and looked down onto the bank where a rotting green cross marked the grave of my pet mouse Batty. I'd loved that mouse and cried and cried when it died on the thirteenth of November 1967.

When I slid my back down the creaking panels of his mews my precious little bird looked up and the sky winked white in the wet lens of his eye. Through it I could see his rusty chestnut iris glowing and the sharp edge of his blue-black pupil and around it a neat yellow ring bristled with minuscule hairs; this in turn was fringed with minute silvery feathers polished with a hint of cinnamon, which delicately swept into his smooth cheeks.

In front of his eye he wore a fringe of fine lashes that curved out around his nostrils and up and into his crown. His eyebrows were lightly washed with cream and fused seamlessly behind his head whilst his cheeks mixed through a faint moustache and into his spiky beard, which along with his crown constantly flexed in synchrony with his moods.

When he was completely comfortable and secure these would be fully fluffed up and out, but when a fly buzzed over he'd momentarily flatten his head and his chin would be shaved of its frilly froth. If a cat crept past his entire body would smoothen, he'd go from teddy bear to emaciated knave, his legs would reappear as he squeezed all the shadows from the cosy ball of feathers, he'd be tight and taut, ready to flap and flee.

His feathers, their volumes and positions spoke precisely of his disposition and I gradually learned to interpret them as I sat for hours pondering his thoughts. He was good at doing nothing, sat on his block or pampered on my ungloved hand where his clean custard-yellow feet shuffled and his blue-black talons pricked the top of my thumb. And I'd wobble my hand to watch him keep his head set still, his stare fixed solid in space, his skull magically locked, his neck stretching and circling as I teased his hidden gyro, and then he'd break it off and bob his head and twitch his wings and shake his tail and his bells would tinkle and I'd catch myself smiling.

He yawned, briefly flashing his mauve mouth, and arched his wings up over his back, fanning his tail; then he stretched first his left leg and wing in unison, then his right, with his toes spread, his wingtips fingered and revealing the rows of dark deckling on the nacreous undersides of his neatly ordered feathers.

When he resettled after a vigorous shake that filled the air with dust I waited for him to fluff up and then I tried to slip my finger into his armpit, fending off his painless pecks until he allowed me to feel the warmth and sniff his mustiness.

And when he was dozy and dangerless I lifted him up and nuzzled into his nape breathing in the dry sawdusty smell, his fairies of down fluttering and tickling my nostrils and he looked straight at me with a slightly pinched frown, bobbed his head and bowed to pick and nibble at his needle-sharp nails.

I turned him to face me and measured the symmetry of the lines of converging arrowhead marks that edged his chest. I looked down his back where his wing blades flicked and folded, where each of six visible feathers were faintly fringed with fawn, and watched thin rays of down fluttering in my breath stream as they frayed from the curled plumes that pillowed above his shoulders. Overall he was the colour of my uncle's sandals, a too-red tan, cheap-looking but with a flush of pink translucence that glimmered on the newness of his plumage. He was perfect.

He cocked his head and traced a line across the sky, sunlight smouldered orange through his nares, he bobbed his head, flinched and then I watched him breathe and tried to breathe in time with his peacefully rising chest until his ashy eyelids pulled up and down and he slept safe inside me as a blackbird spooled out its languid flutery, as its rival clacked and creaked and chinked and my cramped arms goosebumped in the shadowy cool and the neighbour brayed at his bed-bound kids. I left him with the mouse squeezed too tightly in his fist.

The Zoo

April 1970

THE EAGERLY ANTICIPATED trip to London Zoo was the usual family fiasco, my father's attempt at rigorous military-style precision confounded by my mother's contempt for planning, organisation and timekeeping. The result was either 'a bloody carve-up' or 'a nice day out spoiled by your father' depending on who was shouting loudest. But for once not even the spectre of civil war could quash the simple thrill of potentially seeing real live otters for the first time ever. It was sunny too.

As a result of a 'balls-up' we arrived at the zoo late. Once through the turnstiles I sprinted straight to the large illustrated map and, ignoring the very existence of lions, tigers, elephants, rhinoceros, gorillas and pandas – in fact the whole of the animal kingdom bar one group of species – I located the otters' enclosure and set off through the colourful throng. Only to be called back whilst they deliberated over the same map, debated a route, decided that my five-year-old sister wanted to see the primates and that we should all go to the toilets first as we were right by them.

So I sulked in the aquarium, moped around the chimpanzees, huffed in the elephant house, despaired at the reptiles, became comatose in the Snowden Aviary – I mean, birds . . . pointless – all the while whining and pining for the otters. Apparently I had to 'learn to wait'. But why? After all I'd been waiting all my life to see a real otter. Didn't a whole lifetime of anticipation qualify for instant otters? Would yet more waiting make me feel better, would the tedious trudging around antelopes at toddler-pace actually enhance my experience? Why are adults all such idiots?

We eventually reached the otters' enclosure sometime after eating, or refusing to eat, our packed lunch: banana and strawberry jam thin white sliced Mother's Pride sandwiches, a flattened square of Battenburg with paper serviette stuck to it and a rock-hard green 'good for you' apple. I drank my orange squash but only when my request for a bottle of Coca-Cola was flatly denied.

But to my horror they were 'not out today'. They were not out from any angle, from any viewpoint; even when I climbed on the wall, certain I could see the tail of a dozing otter, and got told off, they remained not out. I circled the muddy log-strewn mound with resolute determination but my parents, or rather my sister they said, was getting bored. My mother went to get an ice cream, my father smoked another cigarette and then started pacing and then, frowning, looked at his watch. Surely this could not be happening, this wasn't a 'carve-up', this was turning into a Khartoum, a Balaclava, a Big Horn. This was going to become an otter Alamo and I was prepared to make it my last stand.

My dad said we would come back later. He said we would first have to find my mum and sister who had 'obviously got bloody lost again'. So we strode around London Zoo, him in front, me trailing, his head turning this way and that, his temper fraying, him getting lost, us tramping past the bird house for the third time, the crowd thinning, closing time approaching, when by pure chance we arrived back at the otters and ... they were out!

There were two and they were clearly agitated, scurrying about frenetically whinnying in a delirious game of tag, generating enough energy to make the whole world move. If they were excited, I was ecstatic – galloping around the perimeter, jumping up to lie on my belly so I could marvel at the world's best animals. At one point they sped beneath me and I nearly burst. They were chasing each other and one, the larger, was biting the other and jumping all over it. Some of this play seemed very aggressive, at one point they turned to face each other chirping anxiously, their mouths gaping, fangs bared. I was entranced. Meanwhile my father had got his prized Super 8 camera out and it was whirring away towards the otters. It was amazing and it was being immortalised, I'd be able to watch it over and over.

After a few minutes my mother arrived, took one brief look at the otters, appeared horrified and said we must leave immediately. It was utterly incomprehensible. But true to form she just stormed off and within a minute my dad was also saying it was time to go. He had one more buzz with the camera and then we left. Worse, the following weekend they told me that the film 'had not come out'.

Years later I was sorting through the cupboard under the stairs where in an old cardboard suitcase I found a bundle of tiny plastic reels spilling ribbons of celluloid amongst some broken Christmas decorations, and alongside it the tatty yellow box that held the projector. I took them up to my bedroom and after no end of fiddling managed to project a postcard-sized image onto a sheet of paper tacked to the boiler cupboard door. The erratic and mealy images revealed short jerky clips of my dead, badly dressed relatives trying to look happy at a wedding, waddling like cartoons on a bleak beach and others of my sister having a strop in a blue tutu, myself wrestling with a plastic alligator in the bath, my dad using a net to catch a coal tit as it fled its nest in the New Forest and finally the otters at London Zoo. The clip ran for about forty seconds and they were copulating furiously throughout. And that explained my mother's reaction, and why we had to leave, and why I wasn't allowed to see it. Sex was simply not allowed.

October 2003

'Of course not. How could I, could we, anyone, be happy as a child? Like every other child I was told what to do, where to go, and when, all the time. What to say, what to wear, what to think, at school, at home, in maths, science, music and games, do this, do that – or more often, don't. I wasn't free to do anything, when I was in I had to be out, when I was out it was time to come in … the only thing I was truly free to do was imagine and most of my time I spent imagining what it was like to be an adult. Doing what I wanted to fucking do. The whole thing was totally humiliating.'

It was another of his typically exasperated rants delivered with the assurance that he was incontrovertibly right, and that it would be inconceivable, laughable to suppose otherwise. She waited … whilst a cloud-bound plane gnawed at the glowing window …

'Were there any points, perhaps individual occasions or moments when you were happy?'

He answered immediately, 'Yes, plenty I suppose, but truly happy? Probably only when I was with my animals.'

'Why did you like being alone?'

'I wasn't on my own, I just said I was with my animals, and I didn't say I liked it,' he spat indignantly. 'I didn't like it, it was that it was the only time when I could feel comfortable. Safe maybe. Of course I didn't like being alone, I resented it, at least initially. I hated being left out and picked on.' He grasped the

arms of the chair. 'But when I got older I realised that I couldn't let it last forever. Not being alone, I mean the feeling of rejection, of insignificance, of ridicule. So I separated, consciously. Then I felt stronger for being an outsider. It was them, all of them – and me, just me.'

'You felt superior?' She posed the question suspecting he wouldn't like it.

'Not superior ... I'm very obviously not superior,' he sighed and sat up a bit, readjusting his symmetry in the chair. 'Empowered. I felt empowered. I was outside their sphere of influence so I was in control. I didn't care so I couldn't lose, they could hit me but they would never beat me. Ultimately I would win.'

'Win what?' is what she could ask. Or couldn't: she clearly couldn't ask a fragile patient, who was probably suicidal because he presumed he had lost everything, what it was he thought he would actually win.

'So you were confident that your self-imposed isolation assured you success?'

'Yes. Or no. Firstly, it wasn't self-imposed. I didn't ask to be rejected, I was just me doing my own stuff, they were the ones who decided that I didn't fit in, in that respect I had no choice. And secondly, winning and success are not the same thing, not in the way that most people visualise or describe success. I didn't, couldn't imagine success, can't, not even now. Success is not something that can ever be reached because things can always be better. That's what my father taught me, and whilst you might think it was a tough lesson, he was right. Spot on. But the winning, yes, it might be hard, be a fight, take a lifetime, but I was confident – yes, because I'd been armed with the one critically important asset that I needed ... complete and inexhaustible dissatisfaction.'

He smiled and made rare brief eye contact with her.

'You see I watched those kids, I saw them settling for stuff. I remember walking home once, just before the beginning of the summer holidays, and one nipper told me that if he got Cs in his report he'd get no bike, if he got Bs he'd get a second-hand bike and if he got As he'd be getting a brand new ten-geared derailleur racing bike from Portswood Cycles. And I went home a week later with my report, and sat with it on my knees in my bedroom, in its envelope, waiting for my father to get back to open it and I knew that I wasn't getting a bike. I'd be getting nothing other than being asked why I hadn't got As in maths and being told to "bloody get on with it". And I saw him in the holidays, that kid, with his bike, it was second-hand from Norris's at the Triangle and he was happy with that.'

He thrust out his arms and sat forward, uncrossing his legs and interlacing his fingers, and then looked right at her and said, 'Happiness, that's it isn't it, that's the big problem ... because it's the same old paradoxical recipe for misery. Over the years I've seen people's cravings for stability yield squalor rather than sparkle, their too-easy contentment give them none of the excitement of a struggle against the odds, none of the allure of being plagued with uncertainty or teased by the appalling option of giving up. Their so-called happiness has turned out to be a promise of emotional and experiential poverty, and that's why it, and contentment, must be avoided at all costs ... and the fuel to assure that is dissatisfaction.'

Another aircraft juddered skyward.

'You think that's all horribly arrogant, probably. But it's not. It's just the brutal reality of it. And if constant dissatisfaction leads to continual unhappiness you're probably wondering why.

Why, how, is being securely unhappy the better option? Well, I'd say, it's better than losing.'

At that point the room plunged suddenly into shade, the brilliant flaring of the net curtains faded and revealed part of the drab garden outside, a fly edging up the pane, the broken latch. After a few seconds of adjustment she could see him more clearly too, despite the dimmer light that had closed the room down, hidden the corners and wiped the highlights from the picture glass and the shiny leaves of the plants.

'You don't get the winning–losing thing do you? I'm probably just not explaining it. Winning is not some game, some result. It's not about getting more points. It's not even about the smug satisfaction of beating someone, anyone, anything. It's actually just about never actually achieving anything. It's all about the trying, the striving, the grinding on towards getting a little bit better. In truth I suppose it's not really about winning at all, it's about not giving up. Because that's when you lose.'

The fly buzzed, stopped, buzzed.

'That's when you take the pills.'

The sun rocked in and out and they sat in silence for the ten minutes until the end of the session and then he left, after straightening the doormat on his way out.

7

One Nil to the Mysterons

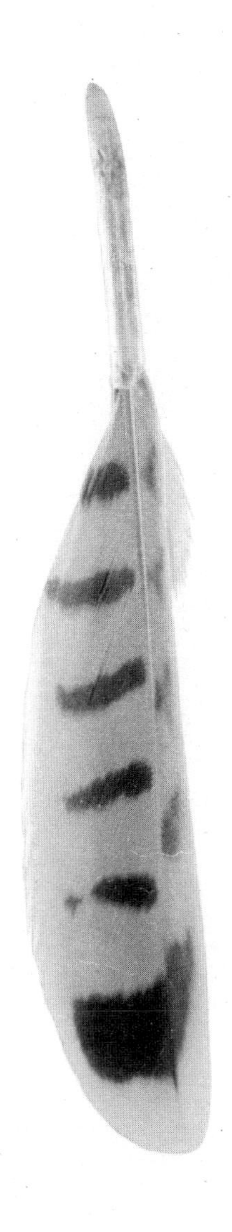

The Hammer

June 1970

It was a Saturday afternoon – colourful, hot, footballs and sun hats, prams and pushchairs and children learning to ride bicycles, winding haphazardly across the smooth tarmac paths, veering onto the grass, falling and waiting for their parents to rush before wailing for the promise of an ice cream. And at the van, queues of short-sleeved dads, smoking and joking, moaning about the missus and Jeff Astle's miss, palming the Brylcreem and then striding back to the paddling pool with everything and one wrong lolly, dripping fingers and everyone's swiping at wasps. The chimps are howling in the zoo but Silvikrin and the summer sale at Owen Owen's and Benson and Hedges Gold are the topics for the mothers sitting, lying on the short scorched turf whilst older kids watch their brothers and sisters splashing in the paddling pool and tell their cousins about *Randall and Hopkirk* because they're not allowed to watch it. Polyester, *Peyton Place*, the three-minute warning.

I had been permitted to bring my net and jars but had been warned that I wouldn't have long at the pond because this was not a tadpoling trip and that there were already more than enough wiggling on my windowsill – those that had successfully hatched from a haul of spawn collected in the spring. We would be having a quick look for newts, and that was it. Then we were all going to the paddling pool and if we were lucky, we would have an ice cream, but not a Heart because they were too expensive.

This schedule was all about my mother and sister because life, apparently, was not 'all about me'. And they liked to sit in the sun or wander hand in hand across the slimy concrete pool, dodging the cigarette ends and lolly sticks and rampaging teenagers and fat Labradors fetching sticks in the ankle-deep water that my dad said was full of wee. Which is why he was sat, shoes tied, socks on, his nose in his pile of library books, only glancing up to scowl at a near miss with a football or to revel in the stupidity of a bloke trying to fly a brand new kite on a day without a lick of wind.

And I was sulking beside this crowded but lifeless body of pointless liquid asking when we could go to the Toad Lake, if we could go to the zoo or was it time to go home yet? Predictably but fortunately, after an hour or so they had something unpleasant to say to each other and my dad and I sloped off for some 'peace and quiet', not forgetting the net and jars.

We began by exploring the corner of the lake behind the island, squinting into the shallows for the shadows of smooth newts. That's how I'd see them; the dark silhouette beneath their thin bodies was easier to spot than their freckled olive backs against the mosaic of sunken oak leaves that carpeted the gravel and gave the water its lovely honey colour. They were tricky to find, especially during the day, much less difficult at night with a torch when many more emerged from their secret caverns amongst the curls and folds of brown, grey and black. Unless I spotted one drifting up for air, snatching a silvery gulp and then flicking furiously back to the bottom to immediately lie as a tiny stick statue, invariably just beyond the reach of my net even when the water was already seeping through the lace holes of my school shoes.

I had left my dad – bored waiting for my turn with the bamboo pole, which I knew wouldn't come until he had actually netted a newt – and was skirting around the edge of a high bank that dropped a steep clay cliff down to the side of the pond.

For a while I'd been listening to a sharp, irregular but consistent clicking sound that appeared to be coming from the shade of some trees that drooped low over the drain that fed the lake. Curious, I scuffed up to the sandy crest of the mound and saw two boys crouched opposite each other on the shiny stone plinth over which the pipe water flowed. They were about nine or ten, a bit bigger than me, and the one on the left had a metal hammer.

I watched them for a while but couldn't see why he was hitting the stone and what the other kid was scraping into a cup from along the sunken edge of their platform. I followed the water's edge round and they fell out of sight behind a bed of stinging nettles, until I crept quietly along the shady path and emerged able to look directly down on them.

I was very careful not to let them hear me arrive. The boy with the cup was scooping up tadpoles from a dizzy soup of charcoal bodies and tails; they had shoaled in the warm shallows to graze a fringe of algae that edged the cement square. He was tipping them onto the centre of the wet stone and the other boy was systematically hitting them with the flat head of the hammer until each load had been exterminated. If some wriggled toward the edge the cup boy would flick them back to be smashed before returning to trawl for more victims.

They were muttering but I couldn't hear any words. They weren't laughing, they didn't seem to be enjoying themselves, they just kept collecting, hammering, killing. And between their knees I could see an inky slick, a black stain, which was slowly trickling into the pond where it fanned out and dissolved over the sunlit silt.

A violent surge of sorrow struck the pit of my stomach. I ran back to my father and told him and then implored and begged him to get them to stop. I didn't cry at the time, nor as we eventually retraced my steps to look down to where they were still mechanically at their murdering, nor as I stood there waiting for him to tell them, to shout at them, to jump down, seize the hammer and fling it into the lake, to grab them and march them to their parents and get them slippered or sent to bed at three in the afternoon. But to my horror he just turned and began walking back towards the rowdy throng around the paddling pool and the ice-cream van and the swings, and I ran alongside him asking why he hadn't done anything, utterly confused. I asked him why they were doing it, traumatised, entreating him to go back, pleading. But he strode on and said nothing.

That evening was a long time coming. I hated those two boys, I wanted to kill them, I wanted to hit them with a hammer and I hated my dad for letting them do it. I hated them long past the TV going off downstairs, the key scraping in the back door lock, the click of the bathroom light switch, the toilet flushing, their bedroom door closing and their voices stopping as the house creaked and cooled. I hated them with tears in my eyes and a hole the size of Tracy Island in my heart.

The Snare

December 1976

I SWEPT THE outside of my numb hand down his soft flank, squeezing coffee-coloured water out of his thin dishevelled fur. He faced away, lips peeled painfully – teeth bared, hard and sharp bones set in the closed crown of his wet jaws. The snare was twisted and tangled, frayed and fur-balled, and cut a fresh pink seam into his neck and his eye whites bled weak blood, a thicker slick of which wound from his ear. I was panting, sweating, soaked and muddied, slumped on the riverbank belching big clouds of breath into the frosty morning and the sun was just yellowing the pines behind me. I'd killed him.

Dick had showed me how to set the snares, how to tie the sliding knot, how to use split willow stems to support the hoop a hand's-height above the path, how to find the runs, check the barbed wire for traces of fur and the thicker black-tipped red hairs that told you that a fox had been crossing there. If there wasn't a fence post to nail the wire loop to then he'd find a heavy log and staple it to that, laying it parallel to the trail. He was the water bailiff on that stretch of Itchen; we'd met when I'd asked his permission to search for a Kestrel's nest, which I'd already trespassed to find in an ivy-covered ash too close to his mill cottage to regularly sneak to. A wily old-school naturalist with a keen eye, he was carefully kind to me, showing me his pinned butterflies and trays of eggs and playfully amused by my incompetence when he let me fire his twelve-bore. He wasn't a fervent shooter but considered it his 'duty' to deal with 'vermin'.

The riverbank footpaths were used by foxes; I'd seen them there, sniffing, trotting, stopping and staring and then bounding off and into the knit of willow and alder. I'd seen their smudged footfalls in the frost-grey grass and smelled their bitterness at the stiles and sleeper bridges that crossed the carriers, that network of ditches that divided the water meadows and often cost a wet foot when my jump wasn't long enough.

When winter began to ache Dick would run a set of snares, fifteen to twenty traps to kill the foxes; he skinned them, nailed up the pelts to dry in the old mill and then wrapped them in stiff brown paper and posted them off. For the pelt of an animal with a good dense coat he'd receive a cheque for twenty pounds.

I'd watched him skin them, slicing with a scalpel from the lower lip down the throat, neck and belly all the way to the base of the tail, and then up the inside of each leg. The skin peeled off the thin bony carcasses easily. He could strip one in five minutes, swearing as fleas hopped up his arms and under the worn cuffs of his rolled sleeves. The tail was extracted using a rod of hazel split in half. One tug revealed just enough of the base to allow the gristly spine to be slipped inside the cleft and then, with each hand clamping the wood and a foot bracing the naked ribs, a swift jerk wrested the chain of bones from the brush of thick fur. Then he'd pick up the bug-eyed corpse, red-marbled and toothed grin, boots of black hair on all its feet, starved, clammy and cold, and throw it into the river where it spun down, a weirdly waltzing log, dancing in the emerald eddies, rolling smoothly into the wrapping weeds, into the twisting veil, a coiling embrace and then across the river's plain of silt, watched by chubb, rudd, roach and dace.

Dick had asked me if I could check the snares the week before. He was away for a night and they needed to be attended to first thing, to be cleared before any fishermen might stumble into them or any of their strangled victims. Most mornings he'd get nothing; some would be knocked down – near misses – and some trodden around – cunning escapes – so I agreed. It would be fine. I rode my bike out in the dark, on the icy roads tramlined by the first few cautious cars and then through the black lanes, breathing the metal morning, the hard cold slap on my cheeks, tears tickling my ears, feet down sliding on the tarmac rink, zizzing soles and hurting hands, two jumpers.

When I reached the river the land was cocooned beneath a metallic gauze, sharp under a bowl of star-spat blue, a pale wash rising in the east and the fields paralysed under a heavy sprinkling of diamond dust. It was just at the moment when the day pinged, when it would be rushed from darkness into assured sunshine, from romantic polar solid to clear wet and brilliant dripping. But it seemed to hold its breath for a moment; crystal-coated in the cruel grip of the ice queen, the night had fashioned a beauty that would shiver exquisitely so very briefly before becoming at once ordinary. As I dropped my bike against a fence embroidered with necklaces of sparkling horsehair the grasses snapped and rustled, a heron croaked in complaint and staggered off, bow-winged, bounding, flaring and landing, brollying up, preening and then staring back as I crunched onto the gravel track and crackled up the sparkling lasso of river.

The first snare was in the fence by the path and hung encrusted with frost. Nearby the second looped into the grass where it had slipped. I straddled the barbed wire and waddled out across the bumpy cow-puddled paddock and ducked through the mangled crush to where the third wire was set between two stumps. It was gone. The staple was there, skewed into the whitened wood, but there was no wire, and the frost was undisturbed so nothing had been caught and struggled free. He must have taken it down. Or maybe it had caught one yesterday.

An angry wren moused through a thatch of sedges in quick nips and whirred off to the split stump of a giant willow where it twitched for a moment, round and brown, hungry, likely doomed. I waited but there was nothing.

Going on up the streamside mallard splashed and teal rose in pairs, levitating in tornadoes of panting wings and slanting away into silhouette, low and level, flying into pinpricks before turning and tilting down and chipping another piece of the morning. Moorhens nodded, barked and scuttled up the glass, treading silver trails, scattering brilliants beneath twists of mist. A silent doily of sixty lapwings butterflyed high over my head, flaring in the corona, pilgrims heading somewhere with a knowledge learned, taught by the strict winter. A few sagged and skidded, but mostly they flew straight and lovely, their pied wings flashing tiny panes of tangerine.

Green sandpiper! Shouting, jerking and down-winged black slanting to the bend and landing, running, bobbing, calling and swiping into the air.

I picked my way along the network of dykes and ditches, gloveless fists in pockets, binoculars bouncing on my chest, striding fast, polishing a path of rime that wound behind me through the reeds and flat grass of the pasture. Out past the old decoy wood with its big round aspens, through fences, over rickety bridges and back onto the puddled path that followed the Itchen up to the six arches of the railway, all the while finding empty hoops of braided wire happily separated from the trails of last night's foxes.

By the time I turned back south the sun smarted a thumb's-breadth over the woods and the frozen steppes had begun to sparkle as frost became dew, each drop flaring miniature spectra before it dripped and all the wonder became just wet. I drove the same heron across the fields in a series of long hops and watched a shrill of finches and buntings glittering over the flashlit stands of dewy thistles and found a covey of thrushes in the hawthorns clustered around the fishermen's picnic area that chattered up to the larches, bare where the river bent and the waters puttered thick and gloopy. There beneath the roots of long-gone oaks pike lay log-like in their frozen palaces puffing their gills, greasy green tigers always waiting to be fast, to make a swirl, to suck some flotsam down to drown.

The snare by the weir had nearly done its cruel job. It was pulled out and furred with wisps of silky grey and one or two scorched black hairs. It had probably got the fox by its front leg but after only a little effort it had been able to pull it free – the vegetation around it was intact, there had been no struggle here. I knelt for some time imagining the scenario unfold, sniffed for any scent but drew a blank, twisted the spent coil between my fingers and laid it down beside the sleeper – defused. Foxes, I mused, were pulled taut between death and a death, kill or killed, the life of a predator. There were three more snares marked on my biro map, all on the river's edge between me, my bike and home and homework.

By the time I'd got halfway along the narrow strip of grass where my shadow was stolen by the shade cast by the rug of conifers overhanging the oxbow, I could see that the log to which the wire had been nailed was gone and when I got to where it had been my heart began pounding.

There in the milky blue cold, through my breath I could see the frost all flattened, the dock and nettles smashed down, the ground green and damp and mud with claw marks. I stood, looked and listened ... the babbling and turning of the river and a robin far off in the pines, nothing more. A rapid scan revealed that whatever was wrapped in that throttling wire wasn't dead, that it had been thrashing about trying to get out and had clearly dragged the log off . . . a badger? The wood was heavy, the mess was big and these animals were tough. I searched for any hairs – nothing, no blood, no sign at all.

I trod carefully further down the path to where it narrowed and ended at the base of a steep, bare clay crescent, cut back by the river and rain-washed smooth, freckled with algae and fringed with long threads of bramble that spilled out of the woods high above. There was no trace of anything here and a badger couldn't scale that slippery face, so I returned to the original spot and rested, slowly calming down.

Then I went back to the main track looking for drag marks, anything, but there was nothing. I lifted my binoculars and scanned down the bank; it was shallow, I'd have seen anything crouched or hiding in the sparse winter vegetation.

So I crossed back over the narrow bridge and walked out onto the promontory of rank scrubland opposite where the snare had been set. The mess of grasses, teasels and briars was hard going, I had to step up to get through, kicking them down as they snagged around my thighs. I got to the end and stood facing the cliff, the torrent racing by, my trousers bristling with thorns, my fingers burning with the cold, and then I saw it.

A head – still, with one wild eye – glaring, a gap, and then the log wagging in the current.

It was a fox and it was in the river at the base of the cliff, right at the apex. It must have been dragged in by the log and then drifted or swum out into midstream only to be swept into the bank where it could only claw at the soil. Desperate scratch marks above it betrayed this. It could have been in the water for fourteen hours, but it had probably been less. As I crossed the bridge again a silky mist was rising from the marbled surface of the stream, stirred by the sun, which was dropping sequins into all its dimples and gilding the surface with lemon. I knelt midway across and reached down with my tingling fingers to the surface and then plunged my hand in up to my wrist.

The cold clamped hard and instantaneously and by the time I'd shaken it and walked to the end of the bank I was sucking at my fingers, biting the pain. I couldn't see the fox from here but I knew where it was: about thirty metres away. I took off my coat and both my jumpers, sat down and pulled off my boots, socks, then my trousers and then rolled into the instantly deep water. A poor swimmer, I thought I'd drift with the current, snatch the log and then float round until I got to a shallower shore. There I'd release the fox and run back to get my clothes, then home and homework.

I tried to hug the bank but as soon as I pushed out the swirling backwash dragged me under. I flailed back to the surface, spun round to face the cliff and began to stroke towards it, gasping and drowning. Now I needed to reach that stake; that dead weight that had anchored the fox in its hell was my only chance of life. I went down again, into the pressing cold, and watched a chain of silvery bubbles rattling through my lips and jiggling frantically up to the air; again I kicked and thrashed and when my head broke the surface and I bit the air I was right alongside the fox, its mouth open, desperate, its feet paddling furiously, its teeth and gums gleaming.

I sank and as I did I saw the bright wire taut in the water, I snatched it and felt it hard in my palm and when I burst back up the fox was pressed by the river against my chest and we were floating free out in the stream, spinning in graceful pirouettes across the enamelled plane of browns and green, frozen, with water in our eyes, mouths and lungs, twirling through the mist, the log against my back riffling, slowing and steering us, a juddering anchor that snagged and rolled us onto a bed of soft silt where for a moment we rested, the fox's cheek stuck to mine, its ear cupping my lobe, its nose bubbling by my eye.

Then our salvation separated us and we exploded apart, both lunging for the bank where the terrible cold seized me with shocking vigour and I began to shake with a frightening violence. The fox was rowing through the mud trying to drag the wire and wood onto the bank but failing, so I reached out with both hands, grasped the log and waded out, towing the choking animal behind me until it was out of the water. I slumped down, spat muck and water and the taste of iron, pinched my shirt and pulled it from my heaving chest, my nails shinning through the paste, and peeled a wig of weed from my neck. Soon I'd be dangerously frozen so I stood and wobbled onto the stony path where I saw my shadow trembling in tune with my spinning mind.

The fox, I had to free it, and quickly, so I edged along the log and discovered the snare was around its neck and that it was crimped and twisted and that there was blood dripping pinkly from its jaws and from the shredded strands of wire. It was crouched now, ears swept back, clamped tight to its visible skull, and it shifted to face me whenever I leaned to examine my options. It wasn't going to let me free it without a fight and I'd need to get to the noose's knot with my frigid fingers; I had no pliers, no knife and no time and I was about two miles from the nearest house and my bike.

I found a fence post and slid it down the snare to pin the fox's head back so I could get to its neck, I knelt on it as it struggled and I palmed its fur apart between my knees. The wire had cut through its skin and had bruised and bloodied the muscles beneath, it had corkscrewed and the knot had jammed tight, the metal threads buckled, frayed and fused. I pressed my soft nails into the coils and, with bleeding quicks bloodying my stiff white fingers, I tried to push them under the tie, but I couldn't, it was too tight.

Again I tried the knot, pushing the brush of sharp ends deep into my thumb, turning and twisting against the throes of the terrified animal. I couldn't undo it. I ran my hand back up the snare to where it was attached to the log and felt and saw the staples driven deep into the hard wood. I pushed back and sat facing the fox. We stared at each other and at our ghastly predicament, we bled and cried, and then I picked up the post and cracked it down on his skull, so hard, so hard he gasped and wheezed and his eyes shivered closed and when he continued to breathe I dragged him back into that bloody river and drowned him. And when the bubbles stopped, I washed him and lifted him out and we lay together side by side everything broken.

The Bird

Saturday 25 October 1975

THE SHARPENED PENCIL made a nice smooth sound as I drew an outline around four small squares and then blocked it in. I blew off the dust and pressed the opaque graph paper flat before Sellotaping it onto the board.

For the last hour I'd been preparing a third facsimile of the previous two tables that I'd meticulously annotated since I'd got him. Beneath blocks entitled 'Name', 'Examination on Arrival', 'Weight on Arrival', 'Defects and Peculiarities' and 'Remarks' there was a long horizontal line of dates from 23 October through to 22 December.

Down the left-hand side was a column of weights, from 5 to 8.5 ounces, and along the bottom a row where I noted what he'd been given to eat that day. Each evening I filled in the little square to form a graph of his weight, measured first thing each morning. Through June and July the little chequered snake was spread between 5.9 and 7 with wide daily variations but after 24 July it descended neatly down to around six ounces where it remained fairly stable. He was fed mainly on beef but the odd HS and S marked in the appropriate space indicated that my air rifle was occasionally picking off house and hedge sparrows in the garden. The mouse-breeding programme had failed. They had all escaped. There were no Ms.

However from the seventh of October the graph had grown a bit tatty again and his weight had begun to drop, sometimes down to 5.7. He'd been odd on that day, he'd pecked and scratched me, lashing out and screaming, and presuming he was over-hungry I'd quickly given him a huge meal. But at the end of the afternoon his weight had actually fallen and he was still in a foul mood. Again I'd fed him generously and again the following morning his weight was down. His mutes had changed too; rather than the typical wet splash of whitewash and beads of black, sometimes with a pepper-mint wash, they were now thick, brown and greasy, like fat slugs.

When I got back at lunchtime on the following day he'd been sick; a series of pellets of undigested beef lay scattered on the sand around his perch. I'd consulted my manual and duly gave him half a Piperazine tablet in case it was worms and some bismuth carbonate, the recommended remedy for a chill, mixed with egg yolk, which he pecked from my finger.

I began bringing him in each evening to keep him warm; he'd sit happily on the back of one of the old kitchen chairs in my bedroom as I did my homework and built an Airfix 1:72 scale Hawker Typhoon Mk. 1b and painted it in desert colours. Unfortunately he'd decided to fly up to the pelmet just as I'd dipped the transfers into a saucer of warm water. He'd lost his footing, tumbled down to the bed and sent the whole kit and crockery flying. The aircraft survived but one of the all-important roundels was lost. I later found it welded to the carpet but despite delicately painting it with warm water and teasing it with tweezers it disintegrated and I had to write to Airfix for a replacement set of insignia.

He seemed better for a while but then began refusing to eat, became sullen, hunched up and shivery and increasingly regurgitated his meals. After school one day I scrambled up into the loft and tore out several large strips of the fibreglass insulating mat my dad had made such a fuss about putting up there. It made me cough and my fingers prickle but I got it down without them knowing and used it to lag his sleeping quarters. I coaxed an old paraffin heater back to life and used this to preheat his bedroom before I put him out each night around nine thirty.

In the mornings I continued to fly him and between days off due to fog and heavy rain and wind he flew better than ever, stooping increasingly frequently, steeply and quickly and not spending so much time up in the 'tramp's tree'. He had days when he'd chase sparrows around the rooftops and once went speck-sized off after some jackdaws, and another time loads of gulls appeared and pestered him until he fought back, ducking and diving around the sideways of Cornwall Road.

But a new problem arose: it was getting light later and this forced me into an increasing number of encounters with dog walkers. One misty morning a black Labrador had bounded up barking and then leapt snapping at him, and on another an idiot with two Dalmatians had allowed them to molest us for ages whilst he bated himself into furious exhaustion.

The window of daylight before I'd need to leave for school was still narrowing and I'd begun skiving first lessons if the weather was good, waiting until we had a semblance of solitude before I cast him up into the air, always feeling an intense thrill at the point that his talons left my glove and he hung for that moment struggling to find his sky before lifting up and banking away, his wings just a thin line, his body a blob until he turned and became my Kestrel, his dark eyes looking back to me, expecting the whistle and the lure.

I noticed two penny-sized 'bald' patches on his crop, I could see his grey skin and the meat he'd eaten churning below, with little clusters of pale bubbles that I could gently move with my worried fingertip. The naked areas grew larger but I couldn't find any feathers that he may have scratched out; maybe he had eaten them.

There were other problems: Mrs Anderton from two doors up had reported us to the RSPCA and an inspector had been round – twice. The first time I'd spotted him, grabbed my bird, nipped over the back fence and across Mrs Slaney's garden and then hidden out in the woods until it was dark and he'd gone. On the second occasion, it was a Saturday and my law-abiding father had given him a grand tour of the aviary and told him absolutely everything about the 'hawk'. After an hour the officer had left duly satisfied that it was getting the best of care but had suggested that I'd need to release it back into the wild as soon as it could fend for itself.

We watched him striding up the hill and turning in to knock on the Andertons' door.

The following week a policeman arrived, presumably at her behest, and he was a lot less interested in birds or their welfare. His parting shot was that we'd need a licence to keep it and if we didn't get one I'd have to let it go. So again I drafted letters to the Home Office and the British Falconers Club and again got tactless, insensitive and unhelpful replies. Then my father made an appointment to see our local MP, Mr Mitchell, and we went along and he actually listened.

I showed him my books, my diaries, my weight graphs and all my best drawings and paintings of Kestrels, minus the one I had given to a girl who sat behind me in French. He'd seemed a bit bemused by it all but agreed to pass on a retrospective application direct to the minister. It didn't work. He sent us back their reply – they couldn't make exceptions, Kestrels were rare birds, it would need to be released or we'd be prosecuted.

Today he weighed six ounces and when I'd released him he'd flown straight up to the big oak, settling amongst its tatty foliage against an eggshell-blue sky. Two twittering finches flipped up and nimbled around him but he paid them no attention; instead his gaze followed an old codger trailing a decrepit dog as he limped around the field in an enormous coat, cautiously planting his stick before he made each dolly-step on the muddy path.

Eventually the bloke doddered out of sight and I called him off. He stooped with a bewildering venom, his wings folded back; heart-shaped with his bastard wings clipped out, he powered earthward and I heard a sword-swipe as I jinked the lure across the grass and behind my back, then he turned very tightly and almost fell on it before I could tug it away again, but after a stall he climbed in a series of tight turns and then jabbed himself towards me again, so fast I worried he would smash into the turf.

But he swept up, flipped over and immediately plunged back, trigger happy. The lure hit me hard in the shin when I jerked it away and, dragging me aside with the rush of his wind, he rowed out, his bells singing as he hissed across the backdrop of the streets, shifting out and up, blending into the red roof tiles before rising gracefully through their aerials into clear air.

I knelt and tossed out his target, waited to see his face and then piped on the whistle. He thrust himself in, disappearing against the dark collage of windows and walls, and then strafed low and fast and silently, his speed pulling the ringers in his bells too tightly to tinkle. Up he went, ricocheting past me, decelerating into a brief flutter over the cherry tree. Some startled starlings fell to earth, before he sculled around and bent low over the field, flicking his wings hard right up to the point that I snatched at the cord and the lure flashed past my cheek.

This was the best he'd ever flown; now he was testing me and I'd disconnected from everything but him and us, twirling at the centre of his converging ellipses, the violent harmony of our sky game, his pure joy of falling, we had finally fused and the world was silent and colourless, only motion, our circular dance, his electron to my nucleus existed. And in the moment of those few minutes, those nine stoops in that shabby arena, I somehow became him, I felt his little thunder, I felt his speeding heart and we sparkled, we shone.

He vomited twice before lunchtime and then only picked at the fresh stewing steak I'd waited outside the butcher's to buy at two o'clock when they'd opened. He ate a little more at six but was sick straight away, as he was again at eight thirty when I gave him some more Piperazine. I put him outside at nine, tied him to his screen perch in the sweaty fug of paraffin and studied him in my weak torch beam. He was beautiful. I was terrified.

The Sparrowhawks

August 1976

THE WOODLAND WAS at its slowest at this time, it hung still and silent as if exhausted by spring's furious spurt, slumped in summer's beery lull, splendid but actually spent. It was painted in sooty green and dun brown and striped, flecked and washed with all the hues between. Only the fiery grenades of cuckoo pint berries sparked in the flowerless groundlights, those few patches that relieved the sucking shadows of the ferny tunnels that straggled into darkness through the forbidding grottos of beech.

A dusty rain that smelled of cardboard chattered in the high canopy, pixelated by millions of emerald gems made from sunlight. Only during these few minutes of washing could their colour come close to matching that luminous lime brilliance unfurled during the first week of May when this great cathedral bloomed. But now the wet whiff of autumn fermented faintly in the cool chambers of these tightly stitched trees and a little death came. Yellow threads ran through its crown and the first fungi pimpled the warm damp nooks and offered their odours to the sweet smell of rot.

He crept in, the vast quiet demanding soft treads, and ducked beneath the lasers of silk, the gleaming threads that strung up the spiders' first naive webs of the season, some pulling like hairs across his cheeks and satisfying his suspicions that only he had trodden these paths that day. Garlands of bryony, their heart-shaped leaves gone to early gold, necklaced down the elder, whose berries dripped winey scent around the drooping cactus heads of burweed.

He followed the hoof-pocked path, the split ovals of roe fired into the clay, heard the faint 'hooeet' of a willow warbler somewhere out on the edge, the maybe call of a goldcrest – high-pitched, lisped into thin air from a perch spun up by the tapering perspectives drawn through trunks, to boughs, to twigs and leaves, gone skyward to spite gravity into a realm beyond our ground-bound minds. And as he stopped, his lungs clogged with the breath of the great green shade, a dragonfly needled back and forth over the path rustling like a paper sprite, its body blinking through golden turns, untroubled by its loneliness. The fallen beech where he settled was furred with moss, tattooed with pennies and coasters of lichen, which were waxy to his stroke and clammy on his side when he lay back and squinted up at the shimmering ceiling star-shot with sky and waited for the hawks to whinny.

He'd been out since February looking for Kestrels, carefully mapping all the nests he found between Southampton and Winchester. The birds at West Horton had not returned to their eyrie in the oak but chosen another across the wet meadow alongside the railway, whose sleepers he had trodden back and forth throughout April and May until he finally sat listening to the squabbling youngsters whilst not revising for his Maths O level.

But in late July, whilst winding his bike uphill through a cool arcade of trees singing 'Starman', he heard a new voice – hawkish, more shrill than a Kestrel and quite piteous, a plaintive yearning. An hour in the woods had given up nothing but a skull on the spoil heap outside a badger sett and an ancient hoard of pornos unfortunately melded into a grubby mottled tube of papier mâché, from which, after an hour of patient peeling, he produced a jigsaw of two women touching each other's breasts. It was worth it – their tanned and toned legs entwined, their lips glowing red and their pink nails plucking at pert nipples, their heads thrown back in glossy ecstasy, when spreadeagled on the compost he had cleared, were astonishingly erotic.

He presumed the birds were sparrowhawks.

On his second visit they'd fledged. They were whining from every corner of the wood, but so softly that they were impossible to track. Eventually the female had flown in and they had all whirled through the spires like a flurry of psychotic bats, wheezing as they whizzed through the boughs in starved pursuit of something dangling from the talloned gibbet of the speeding adult.

One of these fugitives had crashed onto an open limb and immediately death-rayed him with its gunsight eyes, mad yellow lights, sulphurous beads bleeding fire from behind a blackened branch, insane rather than curious in response to its first view of a human. For a moment he felt the perverse thrill of having being found by a killer but then it had panicked and dashed back into the vast hall of trees. He shifted slightly and listened to the bandit posse doing unspeakable things to their prey, shrouded behind the great cloak of leaves. Then they fell ominously silent.

After an hour and a half, he'd conversed with a huge hoverfly, been surprised by a melancholy owl and had a great tit nearly land on him, but heard nothing more of the sparrowhawks. It was seven fifteen; sheets of fragrant light were beaming through the cage of trunks, sharpening the shadows, and in their spotlight things became less ordinary.

Nightshade appeared next to him nodding with its velvety purple flowers, a moth's wing twirled on a strand of silk, spider-strung bunting, fluttering orange spots; a harvestman's pill hung in a basket of glistening wire limbs, quivering down a sunsetting strip of achnied bark and shaking back into the shade unaware of its turn in the spotlight.

He rolled off the log and ducked back to the path, dragging his shins through a no-man's-land of bramble until they tingled and he ouched out loud. His bike was still there and then— He thinks he sees gunmetal and orange, a mesh of rust on cream, he thinks he hears the slicing of air ... maybe glimpses the diamond flash from an eye. He doesn't see legs or feet, and he feels a rush but before he can react, before gravity sets his falling foot to earth, it's gone, bent around the bank in a flaring blur and gone, just the fading rattle of a blackbird hanging in the glade and then gone.

Immediately nothing of this instant is tangible, there's so little to recall that he imagines that he imagined it. It's more of a feeling than anything real – just a fleeting sense that some pulse of life had singed the air. He felt for some fraction of a second a bird fly through him and in that moment he learns more of that bird than he'll ever learn in a lifetime of loving it.

The Punks

March 1977

MIKE WILLIAMS HAD snogged Susan Davies on an armchair in the conservatory. Dave Thomas had put his hand up Lisa Wright's skirt in the corner of the lounge. And John Hughes had groped Karen

Edwards whilst dancing to 'If You Leave Me Now' by Chicago. And no one knew exactly what Rob Green and Kim Wood had been doing in his mum's bedroom for forty minutes, but the consensus was that they'd actually shagged because 'Greeny' had shown Williams a packet of Durex he'd bought and had in his pocket.

Mark Lewis had necked a whole bottle of Bacardi and passed out, the police had turned up because someone three doors down had complained about the noise and Richard Clarke's dad had arrived to drag him out at half past midnight. It had finally ended when Green's mum returned at ten o'clock on Sunday morning.

It was amazing because they all still felt sick now on Monday and Steve Hill still had a splitting headache. Someone had stolen Green's ELO album. And someone had spewed up in the bath and outside the garage. A girl's coat had been left behind but no one knew whose it was. Williams had borrowed his sister's camera and the photos would be back on Friday. He'd skived first lesson to take them all the way over to the chemist in Bitterne because they were 'so bad' that no one wanted them seen by anyone who might know their parents.

I was crouched over my desk, head down, doodling an unwinding viper onto the back of my rough book. It was mythical, edged with hooked scales and frilled with a spiky ruff of hideous whiskers. It would build up all week – them discussing who was having a party, who was going and more importantly who wasn't going, either because they were not invited or because it wasn't cool enough for them to bother with.

I expanded the organism, filling in an armoured skin that reached around the spine onto the coarse grey cover. More of them bundled in and their gossiping got louder. It was a brag-a-thon, a bitching match between rival cliques. The lot blustering around me were the elite, the trendy, fashionable tribe who would only mingle at jamborees with their own caste. Occasionally I'd listen to them inviting outsiders but only if some act of bravado had sufficiently elevated the minion's apparent status.

The boys were creased up, falling about, they kicked my chair twice as I biro-ed two long skeletal fangs up the page. They were all jabbering but no one was listening. Then one of them leaned back on the front of my desk, nudging all my books out of line. Another knock, but he didn't look at me – I wasn't there. I thought about going out to the loo for a wee I didn't need but instead cowered on my elbow, hid my face and overtly concentrated on my drawing. The monster's mouth now encircled my neatly written name and I began to arm it with an array of dagger-like teeth.

This bout of histrionics was the result of a weekend that had seen simultaneous smart-set parties. Friday's book break between double-double maths had seen a gruesome swaggering of allegiances – the two-faced back-stabbing had reached a frenzy and there had been a bit of a scrap. Mr Crosby had broken it up and accused them of being 'trite and pathetic' and told them to save their 'playground spats over infantile vanities' till the weekend. But as soon as he left they were back on to what the music would be, how good so and so's stereo was, whether their parents would be staying away and who would be bringing what to drink. And it was every weekend now and if it wasn't a party then Mondays were all about what had happened at the Top Rank on Saturday night or the ice rink on Sunday afternoon.

This particular orgy had been organised by someone who used to speak to me. Now they were having hysterics with their back to my face whilst I tried to de-exist. Not that it was possible as he spun around and, unable to stifle his maniacal glee, began showering me with anecdotes culled from those they'd all been repeatedly revelling in for the last ten minutes. There was spittle on his lip. A drop of it landed on the eye of the toxic reptile that now tattooed my book.

I watched the dark mote of gob gradually lighten as it dried and couldn't focus on the lesson. I needed to. I was shit at physics and my mock result was only just a C but the clocks had gone forward so I was tired. And the weather had changed. It had been mild and sunny yesterday and I'd found a blackbird on eggs and two freshly lined song thrushes and caught two common lizards amongst the brick rubble behind the Pitch and Putt. But now it was colder and as the final bell had gone it had begun to pelt with rain and I hadn't bothered with a coat so I was going to get drenched traipsing home. I dawdled at my desk pretending to be finishing something in my exercise book whilst I waited for them all to leave.

Finally the door clicked shut and locked all their happy loudness out in the corridor. I could hear Greenaway, the teacher, packing up in the prep room and then a burst of delirium as he opened and closed its door and set off with his satchel towards his bike. He'd be getting a soaking too. I sagged onto the table and sighed. I didn't want to be there, I didn't want to go home, I didn't want to be anywhere. I didn't want to be me.

A familiar hollow ache caved in my chest and I slouched in my woe, watching the trees tussling with the pasty sky. When the last hollers had echoed and fled through the dimmed precincts of the school I finally sat up, tossed my books into my carrier bag and ambled towards the door. I hated them, I hated it, I hated me. Really hated, to the point where for the first time I felt physically violent, like I wanted to hurt everything, break it, burn it, smash it up.

He looked like Bela Lugosi, greased-back hair, an ashen-faced black-eyed vampire. He was squinting and scornful, his lips cracked apart, his nose slightly wrinkled, and he was pressing a large slice of torn meat into one of his eyes, the rest of it flopping down over his cheek and onto his jacket. In the lower corner a text read 'OLD WAVE MEATS NEW WAVE', in the upper in red, 'new MUSICAL EXPRESS'. One of them had left the scruffy magazine on their desk and I was instantly drawn to the tormented ghoul on the cover.

I licked my finger, turned the page and found the charts. 'Chanson d'Amour' by Manhattan Transfer was at number one. I'd heard it on the radio the night before, on the hit parade rundown as I was sat on my bed doing my maths homework, whilst they were all at their parties. I hated it. I scanned down the list ... 'Rockaria' by ELO. As I'd been informed all day, they'd been playing that on Green's parents Bang and Olufsen really loud and it was, they said, 'amazing'. I hated it. One of the girls had taken her top off and sung 'get back inside me Romeo' in time to Mr Big's single, currently at number seven. It was shit, I hated it.

I turned another page: news of Slade going on tour and a headline that read 'SEX PISTOLS FACE EXILE'. I skim-read this story about a group who were banned from playing anywhere in the UK and might have to perform overseas. There was news of some other groups: the Clash, the Damned and the Stranglers, great names.

Then a page about Marc Bolan's comeback tour with the same Damned. I'd liked T. Rex, I'd had a poster of Bolan pouting amongst a feather boa that my father despised, calling him a 'poofter'. But then almost all the performers on *Top of the Pops* were 'poofters' or 'druggies' according to my parents.

I folded the paper in half, stuffed it into my Wavy Line bag and later that evening teased its sodden pages apart on my bed. I read about Johnny Rotten, Sid Vicious and Captain Sensible and a new wave of music the writers in the paper seemed to want to dismiss but always ended up describing as sort of exciting. It was called punk rock.

I'd read through that *NME* several times before a week later I found the *Damned Damned Damned* LP in Woolworths and the Clash's 'White Riot' single in HMV. I lifted the purple plastic lid on the purple PYE record player my mum had got from the catalogue when the radiogram had packed up, delicately positioned the shiny black disc and lowered the stylus. I watched the rainbows see-sawing on the glistening grooves, listened to the sausage hit the frying pan, and then the bass, the screams and then the thump and ripping and slashing guitars and then 'Neat, Neat, Neat' shouted repeatedly amidst the crackling row. I twisted the volume knob as far to the right as it would go, angled the speakers towards my chair and sat down to admire the cake-splattered culprits on the cover.

It was incendiary, explosive – something went off. Somewhere in my head cells fizzed and fired little shocks round my body, which surged and swelled with a desire to scream at the top of its voice. The speeding rants were over far too quickly so I reset the needle and closed my eyes. And again, and again.

Within a week my mother had torn the needle arm from the record player and slammed the whole thing onto the rug after five or six loud and distorted renditions of 'New Rose'. And in the howling silence after she'd ripped the wrong plug from the wall and stamped back into the front room I felt better than I had done for as long as I could remember. I stood shaking over the humming heap of wreckage, so angry yet so elated, and as my dumbfounded sister appeared, strangely dark-eyed, in the flickering misery of the kitchen strip-light with the tap still drip, drip, dripping, I announced, 'I'm going to destroy them, every – last – one of them.'

February 2004

'What happened after that?'

He sighed and without loosening his gaze from the point on the floor at which he always stared he whispered, 'Well, it wasn't good.'

And after a pause added, 'It was not good at all ... it was very bad.'

'How was it bad?' she asked, watching the shadows of the nets blending their patterns with the carpet as they swayed in the draught, his face in shadow, his legs stretched out, crossed, his body slumped into the crotch of the winged chair, the cushions on the floor where he always dropped them before sitting down.

'What happened immediately I can't really recall very well. I do remember being put in bed and lying there hearing the TV and I remember them coming in later and me asking them where they'd put my bird and them not answering. I was lying there, I felt pinned to the bed, my head stuck to the pillow, I was paralysed, I couldn't look up to look around the room, I remember imagining that I had rigor mortis and later, months later realised I'd been too frightened to move, not even my hands. My body just lay there, trying to be dead.

'Then I must have slept because I remember waking up in the dark, not moving, just waiting to know that it was the right time. When it was I got up and got dressed quietly, snuck downstairs and went outside. It was dark and a bit cold, not raining though.

I went into the shed and got my bike out and then back in again and up to a huge set of drawers that were in there, the apple drawers. And I pulled open the bottom one and he was there, wrapped in an old curtain. What was really weird was that I found him straight away in all that chaos. The shed was always such a bloody mess, but I seemed to go straight to where he was without even thinking about it. I didn't even unroll the material to see if he was really there, I just knew it .'

He paused and sat up in the chair.

'Anyway, I put the bundle in my bag and set off to the farm. I don't really remember getting there but it had got light by the time I reached the tree and I walked up and down, along the fence until I had found the spot directly beneath the nest, which was obviously still there. Then I dug a hole with a stick and I buried him. I remember his stiffness, his lifelessness. He was still so lovely, everything was still there, so beautiful ... but dead, and putting him in the hole and then covering him with the dirt seemed such a wrong thing to do to someone so special. It wouldn't work in my mind. You see, I knew he was dead, but at the same time he couldn't be.'

He sighed again and placed his arms down the sides of the chair, made himself symmetrical and then said quietly, 'Nothing that perfect, nothing that loved could be dead. That wouldn't work. It was impossible.' He crossed his legs, broke the symmetry.

She too moved herself, straightened her cardigan, cleared her throat, swallowed, held her own hands and said, 'But he was dead ... '

'Yes ... yes ... he was. Obviously. He was dead. Everything was dead. Nothing could be alive then, everything alive should have been dead too. I wanted the whole world to stop, to completely

stop, everything stop, or be dead. That was the only solution I could see but instead it was just carrying on. And that made me dizzy, I couldn't work that out either. Everything had stopped … and yet … everything was carrying on.

'I didn't stay there long afterwards, I didn't know what to do, I just went home because I didn't know what to do and I don't remember what happened all day, just that by the end of it I couldn't speak.

'And when I woke up on the Monday I still couldn't speak. My brain seemed to want to say the words, I think, or at least it was whirling around trying to, but even if I opened my mouth, which I tried doing, nothing would come out. I remember it being a bit scary, but not much, because by then the mess had started up and I'd lost touch with everything, I couldn't touch the sides, I'd gone, I was out of reach of everything, I was independent, outside looking in on myself. I could see myself from outside which was not good, I was like some sort of … I don't know, but I knew it was bad. And then, do you know what happened … they sent me to school.

'It was a disaster. The geeky outcast now couldn't talk, well, yippee … I just got torn to shreds, not that I cared or even felt it, but the result was pretty predictable – there was never, ever going to be a way back. Of course I couldn't explain anything, couldn't plead or tell them anything because I just couldn't make any words come out. And it went on for about ten days, longer than a week, I don't remember how long. I skived a few days, went and sat in an old shed up by the river, shivered in there, sat there on some wood, watching the rain drip, the wind blow the grass, it would take ages for hours to pass in the freezing cold before I went home and went to my room and went to bed. And

still they sent me back. I didn't fight it at the time, it was all just lost in the bigger mess, but now, I mean forever, I'll never understand that. No one said anything. No one said anything at all, it was as if the whole summer, as if he … everything, just hadn't existed. No one – said – anything. Ever.'

After about five minutes of silence he continued, 'It was like it had snowed, really heavily, that's how it felt. That clogged, flattened, crushed, smothering feeling you get in heavy snow. Deadened. The strange thing is that I can't remember so much from that time, about what happened next. I remember resenting naff things like Mud singing "It's going to be lonely this Christmas" and hating them, it, hating everything and not wanting to do anything. I didn't go out for ages, not until maybe February, maybe March. There didn't seem any point because I couldn't see that there would be anything out there for me, in the countryside I mean. Normally I remember everything, I remember every detail about everything, but I don't remember that. Maybe I didn't even know it at the time, I don't know, but it doesn't exist.

'Probably worst of all was that a lot of the time I wasn't inside me, I was disconnected from myself, I couldn't properly control my movement or speech. I had to tell myself, direct myself to do things and then watch myself doing them from outside of my own body. And all the colour had gone, everything looked pasty, pale, pinky. It wasn't like that twenty-four hours a day, it was generally okay if I was distracted, but if I wasn't specifically doing something then it would start and not stop and it was scary. I thought I'd got brain damage, or gone completely mental, because it doesn't rationally or physically make any sense. Obviously you can't see or watch yourself from outside,

as if you've got the viewpoint of someone sat, stood behind you. I mean how does that work, physically, mentally? It's like some virtual reality thing. Anyway it went on for quite a few months, I remember it happening all the way through a maths exam. In the end it wasn't happening any more. Not then, it's happened again since, but not then.'

'Depersonalisation ...'

'Yes, that's it, depersonalisation,' he replied, 'I read about it once.'

'Brought on through trauma, normally during childhood.'

'Yeah, or shell shock. I knew a bloke with shell shock when I was a kid ... but I never thought I'd be like him.'

He sat motionless for a minute.

'I don't know how I got over it. I didn't, haven't got over it. I still can't talk about it without crying, or wanting to and having to stop myself. Like everything else that went wrong afterwards it just got put away to deal with later, only it never got dealt with. Every year for twelve years on every sixth of December I went to his grave, cycled or drove out there, stood there in the grey, in the frost, rain, mainly rain, I stood there and tried to work it out. Looked at that patch of ground in a field, under a tree, with nettles and grass on it. I stood there trying to give the whole thing some kind of sensible context. But I couldn't get to that point, I gave up, it was never going to happen. I think too many things broke in that moment, things that couldn't ever be mended.'

8

Tyrannosaurus Dreams

The Bird

Saturday 6 December 1975

I WRAPPED MY thumb around the top of the polished barrel, breathed out through my nose, waited for the sight to steady and, listening to my heart, delicately squeezed the trigger. The .22 Webley Hawk Mk. 2 air rifle recoiled with a loud crack that dulled in a fraction of a second to a huffed thud and conjured a tiny bloom of blue smoke as the pellet cleaned the rifling of gun oil. I very rarely missed.

Shooting was the only thing that I was really skilled at. I'd use Mint Imperials as targets, they glowed white in the woods where I shot. I'd line them up on the leaf litter, walk twenty or thirty yards away and then take aim. When I hit one I'd retreat another five paces and try again.

My record was fifty yards using a telescopic sight and around thirty with open sights. It was a state of mind thing, control, discipline, confidence and the ability to regulate my breathing and slow my heart so I could shoot between its beats. In the idiotic Careers Advice interview they'd asked what I was best at. I should have told them I wanted to become an assassin.

Today, firing from the dining room window I had to use open sights, the fine cross hairs inside my scope had got broken whilst I was trying to clean out some dust. I'd been lurking at least an hour, crouched in the damp coffin draught of the window, staring at a constellation of white breadcrumbs laid out at the top of the path to tempt the local sparrows. But after a summer of being blasted they'd become very wary, twittering nervously in the highest twigs of the pear and rowan trees above the aviary, all talking about the gun, all flitting off at the merest twitch of a curtain and taking ages to muster again in sufficient numbers to summon any rash courage.

I desperately needed one; a diet of raw steak wasn't any good, it lacked essential roughage, the bones and feathers that led him to regurgitate a pellet, clearing his crop of mucus, and his crop was not right. It had begun to bulge, to balloon into a grotesque lump after he had eaten, ruffling his breast feathers and propping up his chin. At its sides the bald patches had expanded too and when I put my ear to this ugly boil I could hear hissing and bubbling. Then after a while he'd vomit and look beautiful again.

I missed. I'd snatched at the shot. I got up, shook the back door open and trudged out to spend five minutes scuffing in the mizzle searching for a dead sparrow I knew wasn't there. Just a couple of fresh fluffy feathers glued to a slimy green paving stone.

I pressed my forehead onto the chicken wire and looked into the aviary. His block leaned from a pool of water, his circle of sand pitted with rain craters and ringed with froth. Paint had peeled from the old doors that formed the back of the structure and the rain had washed it into little deltas of white and blue and grey flecks over the concrete and, when I pushed with my cheek, the whole front of the caging yielded and a puddle of water spilled from the makeshift plastic cover I'd nailed to the roof and drenched my shoulder.

This squalid cell seemed so appallingly sad and the summer when it had glowed so brightly and happily so very distant. It was dank, dismal. He was indoors, where he'd been since Thursday, shitting his foul waxy mutes over an old chair, hunched, occasionally shivering, sometimes momentarily well, but not eating and rapidly losing weight.

The graph displayed that tapestry of misery with a cruel clarity. From the seventh of November the line of neat grey squares initially wavered but soon fell steadily from 6.6 to yesterday's 4.5 ounces. He'd been sick nearly every day since the end of October, sometimes he had kept no food down at all. I'd been giving him the bismuth, the Piperazine and some other tablets a specialist vet had posted to me. Most evenings I offered him egg yolk, which he had taken to nibbling from the old tadpole spoon. I'd also been making an infusion of rhubarb, boiling up a bitter brew as directed on page 125 as the cure for inflammation of the crop.

I had sent some of his horrible mutes for analysis at the vet's lab but they had arrived crushed. I'd sent a second batch but not heard anything yet. Those droppings were stodgy now, blood-coloured and contained little beads of fat, and they either plopped in a splash of stinky fluid or congealed on the side of his block. I'd picked through them with a compass point looking for worms, anything, but found nothing – and worse, if I didn't clear them away he would hop down and eat them.

I'd continued to fly him whenever the weather or dog walkers allowed, although by the twelfth of November he had begun to tire easily, landing on my head or my shoulder as soon as I cast him into the air. He'd last flown on the twenty-second; he rose up to his perch on the tramp's oak tree, came off when I'd called him, floated around and put in one good stoop down onto the lure. After that I just got up and walked with him; we found a quiet spot on the side of the path where I used to watch the fox cubs and he'd snuggle up to my chest and follow the antics of small birds that darted from the alders to the brambles and back.

If it was rough I just sat with him inside, hunched over the fire listening to the hissing gas and the grating as the heat shifted the bricks, to the rain, the milk arriving and then to their radio alarm. They'd blunder down with the light on, banging around, and my sister would patter in with her bundle of clothes, climbing into our fire-space to try and get warm and get dressed.

They'd cluster round the table gobbling Rice Krispies, slurping tea and fanning the smoke off their carbonised toast. The paper would clatter through the door and my dad would stride out to get it and hide in it, and then fold it and leave. And then my mum would be rushing everything and get irritable and my sister would want a lift and then they'd be gone and it would be light outside; and we'd sit until I was late and I'd click the fire off and wait for it to un-creak and for the chrome to pop before I'd put him outside and run to school and sit alone in the classroom listening to the muffled deliberations of the assembly until they all poured in with their rubbish.

By the end of the month I was exhausted. I'd be nursing him until ten or ten thirty and then set my alarm at three to check he was okay. I'd sleep in my clothes to save pulling them on in the freezing cold, creep out with my torch, pick up his sicked-up supper, toss it over the fence, before scampering back to bed with my coat on and resetting the trusty Westclox for six thirty. I never failed him, never slept in, never didn't get up. But since Thursday he had hardly eaten and yesterday we had taken him in a box to the local vet who prescribed kaolin and glucose solution. This morning I couldn't mark his weight on the graph, the scale didn't go that low. I'd had to write the figure 4.1 instead of filling in a square.

My nose wouldn't stop running as I slumped head in hands behind the flotsam of wet washing drooped around the fire on a ring of chairs, rising every now and again to check for sparrows until the afternoon crumbled and my dad and sister returned from my gran's. A couple of times I heard his bells rustling upstairs and once the splat of his shit hitting the papered floor. When I went in he was huddled up, his head pulled deep into his chest, his wings drooping onto the chair. He squinted at me and then closed his eyes.

Lying on the bed I watched him sleeping, his head twisting sluggishly round and then correcting itself, twisting, correcting, his eyelids fluttering, his brows twitching – he was dreaming. He made several barely audible squeaks and then struck out half-heartedly with his anaemic foot.

He was somewhere else, not in this choking hot room on a grotty chair, perched above a fetid porridge of dirty droppings, huddled in the half light with a mucky chin and grubby chest and a tail whose edges were frayed and dry and stained. He was out, tussling with currents of cool air, soaring through clouds and spinning after swallows, floating weightless, upswinging and falling away, he was strung on a glistening wind, crashing through a rainbow, he was flying for the sheer joy of flying. He was free.

We had tea whilst Jim fixed it and then at six thirty I plodded upstairs with my dad to feed him some glucose solution. I took a towel, wrapped it around his wings and sat on the edge of the bed with him laid across my lap. My dad used a syringe to draw about 10ml of the clear sticky fluid from the jar. Then he crouched down alongside me, eased open his beak and began to slowly compress the plunger. At first he swallowed it, but then he began to struggle. I told my dad to take it slowly so he pulled the syringe away. He gasped once, then his chest heaved and his eyes began to flicker.

Then I heard my mum's voice prompting me to kiss him goodbye. And then he died.

It was strange; I suddenly noticed we were all there, me on the bed holding the swaddled bird, my father kneeling in attendance, my mother stooping over his shoulder wringing her hands and my sister standing separately, gaunt and puzzled. I could see it as if I was looking from outside and it was like some surreal nativity scene. Except the baby Jesus had just fucking died. And for just this one wretched, lingering moment I could taste their horror. Smell their shock and their fear. I could sense the terror detonating within them, the knowing, like they'd just seen an atom bomb fall and they were grappling with the realisation that the consequences would be total loss. Total. Loss. But in a second it was gone, my father rolled the bird in the towel and took it out, my mother wrestled me into bed, turned out the light and closed the door. Gone. Darkness. Confusion.

The Nightmare

November 1976

IMMENSE TREES WERE crashing in an onrushing tide of mangling branches, a cascade of explosive violence that jammed the air with a choking confetti of debris. He was stumbling, clawing his way through, pressed forward by the force of a racing storm of juddering destruction, with bloodied hands and bruised everything, tumbling through a squall of dusty moss and leaves, his mouth dry, coughing against desperate gasps of thin air, his chest heaving as he hurdled falling trunks and thrashed through the basketry of thorny jungle.

Lightning-bright sunspots tore through the humid curtain of the trembling canopy and blinded him as he staggered through sucking mud and ricocheted between the smashed maze of vegetation.

Behind him hell had opened and its vile voice belched deafening roars and the sickening moans of ripped and torn timber, cracking and splintering, sharp and loud and relentless. And he was tiring, he was floundering, cramped and scrabbling hand over hand through a soft swamp, losing traction, energy, faltering on his feet trying to fall forwards.

So he kicked again, put the pain aside and stoically dragged himself over an enormous mound of rotting wood before the flood of disintegration overran him, punching him into the air and hard into the crumbling crown of a gutted tree. A blizzard of leaves swirled and then slowed in the shocking decay of noise; behind him he heard the whole world groaning, aching as the circus of atrocity staggered to a standstill and in that lull he rolled over on his back, ripped, winded, wrecked.

And it was there, panting, the hammock of its chest heaving, dripping a soup of litter and slowly lowering its massive head. It shook off a slurry of the forest in a twisting ripple that ran from its snout to the distant tip of its curling tail and snorted, swallowed and then cracked open its jaws to reveal a cage of wet dagger-like teeth. A ripe meaty breath spewed over a flexing tongue, tossing gobs of foamy spittle through a crowd of flies spinning around its jowls and crowding on an oozing wound beneath its gleaming eye. Then it stilled, its whole body tightened in a seizure of concentration, its jaw dropped open and it lunged forward in a horrifying quake of volcanic pressure.

The drawing pins had rusted in the holes at each corner of the picture. He licked the stains, cringed at the twang of metal and rubbed away the brown with his thumb. The two photos had been on his wall since that awesome morning nine years ago when he'd eased them out of the envelope from Pinewood Studios. They were black and white lobby cards from *One Million Years B.C.* He'd paid for them with a postal order and he'd loved and looked at them for ages. But Ray Harryhausen's dinosaurs – the Allosaurus in the Shell tribe's camp, snarling at the spear-wielding and bearded Tumak, and the Ceratosaurus confronting the Triceratops in the stony desert – were now officially extinct and had to be retired.

He'd always been vexed by inaccuracies in the film – Ceratosaurs were Jurassic dinosaurs, they had all died out before Triceratops had evolved and although they did live at the same time as Allosaurs they were smaller. In the movie they had it the other way round. There were other issues too. The animated creatures were dull and un-patterned, but why, when large living predators like tigers and leopards were striped and spotted?

In fact why were all the dinosaurs in films and books and the models at the museum grey or brown or olive green? Just because adult alligators, or crocodiles or Komodo dragons were. But what about all the other reptiles? All the banded, streaked, freckled and spotted lizards and snakes? It didn't make sense. Nor did the idea that they all lumbered around like fat prehistoric zombies – obviously the carnivores would have to be able to chase, catch and kill their prey.

Well, finally scientists were thinking that they did, as he'd enthusiastically learned from last night's *Horizon* documentary. It was his favourite TV programme, full of facts, and 'The Hot-Blooded Dinosaurs' was a brilliant one. A group of scientists believed that dinosaurs were not actually cold-blooded like modern reptiles. It was a controversial theory but they had loads of evidence. Analysis of their skeletons actually showed that they stood upright and walked on two legs; this meant they could be more active and had to be able to use more energy, which in turn meant they had to be hot-blooded. And to prove it their leg bones had more blood vessels in them than previously thought. And what's more, they weren't even extinct – birds were the descendants of dinosaurs. It was all to do with their hips and legs.

He put the pictures into a box on top of some babyish books about animals and the letter he'd got from *Blue Peter* informing him that they couldn't disclose where Dartford warblers lived because they were too rare and specially protected. In the corner was his tail-dragging plastic Tyrannosaur; it had formerly been grey but he'd painted it orange with purple spots, maybe a little more Bolan than Cretaceous.

The box was one of a collection scrupulously stacked beneath his bed that held the history of his life in objects – catalogued, filed and preserved for possible posterity. He found it impossible to throw things out that had once held value, there were tomato boxes full of old toys getting dusty in the loft and others bursting with all his drawings and paintings.

He reached onto the bed and added *A Gun For Dinosaur* to the archive, a book he'd been given by his old art teacher Mr Hann. It was a story about a time-travelling hunting safari run by a Mr Reginald Rivers and his idiotic clients whose antics get them killed by the wounded reptile. He liked the plot, it had funded plenty of his fantasies – him and Raquel Welch living in a cave, fishing, gathering fruits, dodging dinosaurs and sleeping under a bed of furs. But he didn't like paperbacks and this one was tatty, and worse he hadn't had it from new. He hated second-hand things. When he grew up, he'd never again have second-hand anything.

Nevertheless, together with the *Horizon* programme it had no doubt seeded his nightmare, but then he often dreamed about T. rex, always had done, because it was the greatest animal that ever lived. It had the three key attributes: it was big, fierce and, crucially, it was extinct. And that meant that neither he nor the scientists would know what it really looked like, what colour it was, how it was marked, how it moved. It was dead but it would continue to evolve as long as they studied it and he lived and dreamed. T. rex would live forever, it was immortal. But he wasn't, so the Tyrannosaur he would uniquely imagine as he grew older would die with him.

As he slid his memorabilia out of view he thought of his own death. He found it strange, it felt somehow improbable. Everything seemed to be organised as if he wouldn't die. People consistently projected ahead, said they'd do things later, tomorrow, next week, next year, when they grow up, get old, retire, as if an inexhaustible amount of later was guaranteed. It was all structured as if life went on: his education, his going to college, university, the course of progress all depended on life's continuation.

But it wasn't like that, the only thing truly guaranteed was death and yet even though it was always so close it was never factored into anyone's plans. Because none of them knew exactly when they were going to die they sort of thought that it would never actually happen. They didn't plan for it. In fact they were terrified of it, so incapable of accepting its finality they had to fabricate nonsensical rubbish like reincarnation, resurrection or an eternal afterlife. His bird wasn't living forever in heaven, it was rotting on the edge of a field, just like they'd all rot or burn, like he'd rot or burn. Maybe tomorrow.

That night he dreamed of Tyrannosaurus again. But this dinosaur was different. It had been transformed and re-emerged in his subconscious as a terrifying dandy, cloaked in a rippling flux of glimmering feathers and bedecked with a fluorescing Mohican of orange and red plumes that bobbed as it jogged and then fell flat as the reptile saw him and lunged into its earth-shattering gallop.

The ritual pursuit of the dream accelerated rapidly into its hopeless cascade of bloodying collisions and panic-ridden chaos. He could feel the dinosaur pressing him towards his predictable death in its massive jaws. And when he collapsed exhausted and rolled over, there it was like an insane predatory peacock, its flanks heaving air through its gaping mouth, swirling a small flock of its own freed feathers whilst wefts of patterned plumes fanned out into an aggressive frill around its head. Then it erected its crest and surged forward to kill him.

But rather than waking as usual, in that shocking instant the forest reappeared and after a moment's confusion he sensed an immense weight and power. He turned his head and swallowed and recognised his point of view as that of the dinosaur. He bowed and saw his great clawed feet planted deep in the moist earth and watched a slurp of his former blood dribble into the stagnant compost. His jowls slapped as he shook his head and he swallowed a heavy gulp of treacly saliva before, from a gut of colossal depth, he punched an almighty roar that quivered his cheeks and clouded the air with a septic mist. Then he shifted and the scale and strength of his body became apparent; his muscles moved freely but with the feeling of a hefty tonnage, he absorbed the lift and bend of his legs, the distant corrective rocking of his stiff tail and the immense force of his footfall, obliterating all as he slowly twisted through the sun-spattered forest.

And then he was running, his sheer mass bulldozing him through the trunks, he could hear the thump of his feet and the explosive spasms as his huge heart pounded in his chest. He had become the Tyrannosaur and he felt the power of his rages escalating to an ecstasy of momentum, he felt that immortality as he ripped through the trees chasing himself down, satisfied by the secure inevitability of his own fatality, he felt at ease with his unquenchable destiny to kill himself, with a confidence born of primal instincts. And as he careened to a shuddering standstill over his fallen self a euphoric rush of destruction woke him and he jumped up to wring the sweat from his sheets, terrified of himself.

The Treasure

July 1974

SHE STOOD PICKING the pebbles from the dash, locked out on a sweltry afternoon, leaning in the doorway under the rose, her bobbed hair smelling of shampoo, in corduroy jeans and espa-drilles with grass-stained toes. Transistor static fuzzed over Trumpton and sporadic cap-gun fire and next door's scraggy lawn was peppered with luminous plastic toys, disconnected and neglected by the disaffected pair of orphans whose snot and scabs stained their yesterday's shirts and socks whilst they made too much noise.

A door barked, a dog slammed, a tired butterfly sagged over some wilted daisies, the yellow beak of a shiny bird dribbled notes from the eaves and a sun hat with a pram, sparkling wheels and clicking heels crossed the road. Everything was burned and bleached, the sunshine was exhausting and in the faded shade, with her fingers smelling of the cream soda she'd spilled, sticky, sweet, she flicked away a stone and it stuck with a tick to the tacky tarmac. She slid onto the step and sighed, a doorbell bing bonged, Avon was calling but she couldn't see anyone arriving anywhere.

Her mum must have popped out, probably to the Castle, a parade of shops next to the pub of that name. Earlier she had gone over to Maria's to play but they had rapidly run out of things to do and then to say. Her new rabbit wasn't friendly. She'd tried to hold it, to stroke it but it had kicked itself free and bolted under their shed. They'd poked it out with a rake but then Maria wouldn't let her even try to hold or stroke it. She said she could get the old rabbit out, but the hutch stank and Doc was backed into a corner on top of a mushy pillow of poo and she'd just had a bath and put clean things on and besides, that rabbit had never been friendly either.

She'd watched its whiskers twitching, it had squinted at her with a mean look, its peeping paws stained brown, a cluster of flies bothering a tear on its ear and when Maria went indoors she'd closed the latch on its cage and dashed round the sideway and back up the hill and home.

She pitched another stone chip into the face of a pansy and skipped round to the back of the house. Through the window she could see a jigsaw of pattern paper lying over a set of cut fabric laid out on the sideboard and two limp swatches hanging from the sewing machine, their edges pinched together in the chromed jaws of the shiny black beast. There was no reel on top, her mother had run out of cotton, which meant she'd have got the bus to Portswood and would be gone at least an hour – longer if she went into Kelsall's to get food as well.

She dragged open the shed door and tiptoed into its webby chaos, found a sponge in a bucket and wiped the top of the kitchen stool, which had been left under the dripping clothes wringer. Then she carried it round to the front garden and used it to climb through the open window. Her knees hurt on the windowsill but as the nets drifted closed behind her the cool stillness of the room felt good. The house was quiet, in the kitchen the tap was still dripping and she could hear the electric meter ticking by the stairs.

She edged up to the Singer – glossy, robust, precise, it was Victorian in appearance and decorated with fiddly scrolls of fine floral gold. She stroked the serial number, ran her finger over the embossed badge and then gently rocked the polished wheel, which felt much heavier than it looked. Trimmed threads wormed and wriggled out of the fringe of cut cloth and offcuts lay like leaves on the table. One, snagged on the frayed veneer, dripped a mesh of glistening strands that sparkled sunlit in a tiny cosmos of dust.

She liked the sewing machine. Sometimes she was allowed to turn the handle but never to feed the fabric past the jumping needle that bounced through the chromed footplate as her mother nimbly steered the seam. She felt the material, a mix of heavily patterned satins already stitched into a bodice of black plush velvet. It was a costume for her dance school's forthcoming annual performance – the Elfin School of Dance, where she had been taught ballet every Wednesday since she was five.

She had wanted to be a dancer, she'd worked hard, tried to concentrate on everything Miss Moore had shrieked at them in those echoey rooms. She'd tiptoed and tottered in circles dressed as a butterfly, loved her first leotard and having her hair scraped into a bun, but now she enjoyed ballroom more – every Saturday in the church hall. She got to dance with boys, she was allowed, but not to go to their parties. She had to lose those invitations.

At the top of the stairs his door was closed. As she opened it the house creaked as if in warning. It was completely ordered and tidy, everything was aligned or equally spaced, his purple bedspread had been smoothed perfectly flat and his beloved Newton's cradle stood centrally and symmetrically on his drawers. He'd painted the fronts white with purple discs to match his preposterous inflatable chair, which stole a significant volume from the small room.

The gecko light was on. The lizard, immobile as ever, hung upside down in its cave, its lashless copper eyes slit with deep blue, the only liquid things in its arid tank.

A mosaic of colourful posters and postcards covered the walls – birds, butterflies, moths, their wings spread elegantly in grids and groups. They looked better than in real life. Real birds were distant, dull and disappointing, butterflies just blurs, but here spread on the white paper they shone bright and bold. He'd marked them precisely with various symbols. The butterflies had a mixture of biro ticks, crosses, dots and dates alongside each drawing and in the lower right-hand corner was a legend that explained that a tick meant that he'd 'Seen Adult', a cross 'Seen Caterpillar', the dot denoted that he'd 'Kept Caterpillar'.

Cooking on the windowsill were his skulls, a large brown feather, a circle of air rifle pellets and the Airfix Spitfire he'd been preoccupied with the previous evening – she could smell the paint and glue. A squadron of other planes hung in formation, strung to the ceiling with an elaborate grid of threads, and a fat fly smashed itself against the window between loud sorties around his purple plastic light shade. There were always flies in his room.

In front of her, set centrally on the bookcase he had made, stood a small blue-grey chest, each of the eight drawers measuring about an inch and a half high and thirteen or fourteen inches across. None of his immaculately ordered clutter rested on top of or leaned against it and despite its modest simplicity it was very clearly the room's altarpiece. It radiated sanctity, it commanded awe, she knew it was a repository of extraordinary value to him and as such it was absolutely terrifying to her. She should kneel or bow but instead she stepped up to it and touched the top handle, above which an absurdly neatly written label read 'Finches, Thrushes, Starling, Robin, Sparrows'. She felt a powerful charge of betrayal and violation shoot up her arm but rather than calmly and respectfully backing away she tucked her hair behind her ears, slowly hooked her finger beneath the handle and pulled.

A slow, dry, woody scraping revealed a pristine bed of flattened cotton wool, flawless and brilliant white, followed gradually by rows of tiny gems, each neatly nestled into the gossamer tray behind tiny typed labels, all intricately regimented with astonishing precision. She'd stopped breathing, her hair flopped down on one side and she flinched. Her head was humming, she was shaking ... what on earth was she doing? This was beyond a transgression, a violation, or some petulant defiance, this was the worst sin imaginable. And yet as if possessed she continued, withdrawing the tray completely, setting it down on the bed and then kneeling, hands clenched and trembling before its treasures.

She could see now that they were beautiful beyond compare. Every unique little capsule of impossible fragility was marked with such natural delicacy and subtlety, all washed and painted with a palette of sublime colours. And each had been patterned inside something free and wild and unaware of its spectacular ability to produce such radiant loveliness.

She studied the rows and columns, compared almost whites with blues and those flushed with rose or cream or grey to those with cinnamon, tan or green. She marvelled at their markings, the speckles and blotches and clots, lines and bands so fine, so delicate, so clearly beyond the scope of any human hand, so obviously randomly applied and yet so incomparably perfect.

The range of shapes and colours and decorations was wonderful: the smoky maroon smudges and hairlines that ringed the crown of the bullfinches, the soft drizzle of orangey-red that mottled the robins, the charcoaled spatter that smothered the maybe grey, maybe white of the house sparrows, the pallid blank blue of the starling and the rich tint of the little dunnock, which looked like a tear dropped from the sky.

One, the reed bunting, was a lovely mellow mix of mauve and warm coffee brown that graded to a thicker ashy cap and was twisted with a ghostly lace of faded squiggles overlaid with a wreath of frail black text, amazing scribbles scrawled within the body of a bird.

She leaned back, sat on her ankles and breathed. The fly bumped electrically up the milky glass and she reached out and gently tipped the drawer up towards her. The display was livid but not lurid, it glowed, but none of its hues were bright or dominant, it was so rich in form and colour and yet quiet, rested, resigned.

The collection itself had a childish charm but also an unsettling edge; it was just too neat, too precise. There was not a trace of carelessness, nothing casual; it was a manifestation of a mind obsessed by intense detail and absolute precision. As her eyes carefully scanned the rows she noticed one of the eggs, a grey wagtail's, was on its back and that cut into it was a minute circular hole through which he had removed the white and yolk. The incision was so small, and the shell so thin, she couldn't imagine how it had been achieved.

Altogether it was remarkable, a jewellery of natural masterpieces in miniature and a small study of the life that flew around their neighbourhood. But for all its wonder the chest was a tomb – everything it contained was dead. How could he be celebrating life by so calculatedly ending it and then revelling in its beauty at the expense of its death? When it was life, the birds, that he loved so fervently, how could he be so spellbound by this hoard of vacuous carcasses?

The third drawer down was labelled 'Owls, Falcons, Hawks, Grebes, Coot'. It held larger eggs, some round, white, one almost diamond-shaped, shiny and washed with a bluish tinge, and a couple of spherical shells heavily dappled and smeared with chocolate brown. As the light fell on one she could see it was cracked, a hairline fracture jiggled across its surface, an almost imperceptible flaw. Another glistened with traces of glossy glue. He'd somehow rebuilt it from tens of tiny fragments. There were some pinprick holes but it must have taken hours.

There were fewer in this tray but in the centre with all the others arranged around it was a rich reddish brown and densely freckled egg. Beneath it the tiny paper label read 'Kestrel'. She knew this was his prize, his sacred favourite, and it was obviously precious in its speckled matt shell, lying there ringed by lesser things.

She touched it, stroked it with the tip of her little finger, barely brushing it but sensing the texture of its surface. She began to rotate it in its snowy cocoon and it felt quite robust so she gently edged it up and out and then cupped it in her hand. It rolled across her palm, her heart leapt and she drew it into her chest until it settled.

Suddenly the realisation dawned – what had she done? She'd trespassed in his room, walked up to the cabinet, opened the drawers, taken them out, picked up his pride and joy, and now she had to get it back in. Kneeling up she craned over the tray, drew a breath, held the Kestrel egg between her trembling thumb and forefinger and began lowering it back into its cotton wool cup. At that moment her mother burst through the front door and her finger punched through the shell, which instantaneously collapsed in a confetti of dry, crispy fragments. The end.

The Present

December 1974

BRITTLE IN THE mist, the surface spun and steadied and circles of lacy white radiated and were spent in the suit-grey channel, and ripples kissed the bank, and downstream a branch swivelled through rolling eddies with silvery curves, and bubbles, and a surge and up onto a bar an otter beached brown, so quick and shiny with a halo of glistening whiskers and gone in a slip beneath and back again, its head bowling over, a liquid log pouring into the breathless cold, into the twiddling roots of the alders.

He saw the scarf of otter bend back, a twist of eelmetal spooling a chain of brilliant fizz, and then its head rise, its arch flex in the flux of river, furling down in a thrill of fluid motion, giddy zip and curling coils of tail in a whirl, a twirl of two, both slurping onto the bank with a yip and a nip and ebbing splashlessly into the pool to bow and sweep away, porpoising arcs in the out-stream, warping the plane into the crouch of the big willows, quiet on the reedy shore.

He gasped. His first wild otters. He could have shouted, sung, but retained just enough composure, distracted by his elbows soaked through his parka, his binoculars wedged into his sternum, the numbness of his toes in his wellingtons, the wooden smell of mud and the soft salty taste of snot that had drooled into his mouth.

A slop, and they were back, one so close on a reef of sand, smacking its mouth on a fish, ivory and floppy, gnawing, rocking its jaws from side to side, its eyes winking, dipping into the current, shaking its frilly moustache, its sepia coat cracking tan, long splits running away from its neck and then shaken into a spiky ruff, teddy bear nose with pink spots and stopped ... frozen still ... staring upriver and then fast back to jerky crunching, grimacing, the pearly flesh pulped headless, snatched from its scattered shoal somewhere in the deadlight of the stream amongst the menacing wrap of weeds where the others turn sleek, fluttering in the flow, steady in the glass, olive backed and scared, flinching darkly in river-fear.

The otter listened and smelled, nibbled abruptly at its flank, rose up, shivered, shook a faint chandelier of spume from its trunk and dissolved. After twenty minutes' more biting cold he crossed the bridge by the island and snuck down the bank to where it had fed. Nothing remained, no footprint, scuff or scales marked the mud, no souvenirs of the moment to gather and store. A pheasant crowed and whirred in the poplars and a party of redwings tumbled into a jumble of hawthorn to bicker over the last few sour and scabby berries; the river chattered in odd pops and smacks.

It was over and at home they would be up by now, already nursing the tensions of Christmas Day. Tiptoeing around the dread of expectation in dressing gowns, sipping tea, turning on the TV too early, plugging the fairy lights into the two-way, clicking the starters on the gas fires, jiggling the thick curtain across the back door on its squealing rail, knocking tinsel from the pelmets, making a proper breakfast, over-winding the musical tree, opening the door and letting all the cold in, all the while the three of them moving in separate spaces.

And beyond there millions more dads were sighing and groaning, tugging on their Christmas cardigans, yelling downstairs to kids buried in drifts of torn paper and getting ready to hate their relatives with Instamatic cameras and sherry and eggnog and party sevens, and millions of mums were in kitchens peeling and basting, shouting at the kids in the lounge with their Six Million Dollar Men, Flight Decks and cap guns, out in their sideways with Tri-angs and Bantels, and screaming upstairs at teenagers with Fidelity, Ferguson, Sharp and Pye record players already scratching sounds from Slade, Mud, Abba and Tina bloody Charles. And at G-Plan tables girls in denim and polished Mary Janes sit pecking on grey and green portable typewriters by Olivetti and Brother.

By mid-morning the mums would be giving dads Brauns with foil heads and dads giving mums gold-plated, seventeen-jewelled shock-protected ladies' watches by Old England, and if it didn't fit, was the wrong colour or size or broke in the first five minutes it didn't matter, because everything came from Argos. Except the single greatest Christmas present in the history of the earth … because Argos didn't sell otters, in pairs, eating dace about ten yards away. He gave up escaping, marched straight back to his bike and raced home to a roast and a row rendered irrelevant by his very own ring of bright water.

The Clash

November 1977

THE HUMMOCKS OF naked sand were huge and bunked in tiers beneath a scaffold of oaks whose boughs drooped lower than their roots over the steep bank, down to where their twigs twirled plaits of flotsam snagged from the river's spate. All the light had been let in, the cold had stripped the canopy and a great litter of papery scraps fidgeted brown and scurried in excited crowds over the ground, crushing themselves into crooks and clotting in cages of bramble.

A decrepit walkway girdled the base of the slope, blackened and fragile, strung together by a net of chicken wire and rusted nails, wobbling treacherously above the deep sweep of the bend, where in the twittering backwash I imagined packs of perch playing havoc with spinning fry. Linking the bunkers was a map of smooth paths swept clean of leaves, some with precipitous drops to the water below. I trod them, dolly-stepping their twists down and following their curls into the woods where they continued to run true and clear, and then back to the giant sett where I sat above the sleeping badgers and flicked sticks into the flow.

Against the black water yellow-yellow leaves wrote autumn in dizzy circles and spoke silently of desolation. It was cold, I had two jumpers on but still felt sickly and my matted woollen gloves offered little respite from the snarling wind that hissed up the bank and scraped the skin from my cheeks.

I'd come out because I had to. I didn't really want to be here, or anywhere and I just couldn't muster the motivation to look for anything. I'd seen a few birds, found a murdered woodpigeon. But I didn't have anywhere else to go. It was either my bedroom in their house or the woods and the fields. I couldn't work out whether I wanted friends or a girlfriend or just wanted to hate them all for having friends and girlfriends. Either way it hurt and although I tried to control the mulling and mauling and mutilating myself, I just couldn't stop it. So I yanked the zip up to my chin, pulled up the hood, tried to stretch the cuffs over my wrists, wrapped my arms over my chest, slumped beneath a knobbly gut of bark that bulged from a burly tree and desperately tried to just sleep the afternoon away.

I stirred to see Dick's Land Rover snorting along the track behind the frieze of poplars where I'd watched those otters a couple of Christmases back. Then I must have dozed for a good while before waking up quickly. About two yards away was a badger. It was almost completely out on the spoil heap, nudging the air and sniffing loudly, posing with a sturdy grace, with its smooth smock of silver grey and fringed flanks, its black legs and white-tipped ears and three stripes pointing at its dirty nose.

The little bear sneezed, twice, and then fell onto its side to scratch its long neck with a calm and satisfying rhythm. Suddenly it flinched and sprang onto its feet, startled, gazing through the girders of oak out across the river and water meadows as still as a tree. Then it slowly smudged back into its bramble-ringed pit and sank with the sound of a shadow. It was the closest I'd ever been to one, but I was conscious that I hadn't really connected with it, that I wasn't as excited as I should be. I waited for five frozen minutes and then scrambled up and stole out, back to my bike. By the time it re-emerged I'd be miles away.

Thump thumping, dark low, gut-punching bass and vague voices that phased and faded like those I'd heard as a kid, tinged with menace, moaning a warning. And I press through the mob, all the peeled eyes and shoving, the sticky smell of plastic lager slopped, milling in a boiled air of frayed desperation, obvious violence, smoke, studs and badges, squinting to see faces in the mucky foyer light, staring at people staring back, at girls dressed in flesh, and with butts choking the urinals, I'm piss-treading and watching my back in leather and chains with an anemone of biro-blue spikey hair.

I'm half scared, twice as excited and relieved to be here because this rebellion has arrived in the nick of time. I dig my way down to the front to exercise my strength of will, wound up with fury to shout, to scream above a noise spawned by a nation's ruptured frustration. Crushed into a suffocating scum of steaming sweat and jostling scowling youth, we watch the roadies stomping around the stage, fiddling with stuff, one-two, one-two-ing into the shiny mics, smacking a tom-tom and making the tiny orange lights on the amps flicker and the speakers cackle whilst fat security wankers patrol about in glossy bomber jackets waiting to bovver someone.

And then the music stops, the lights go out and everyone cheers and pushes and I can't breathe and silhouettes of the Clash run out and with a click and a hum it all explodes into a frenzy of dizzying churning as the stage ignites with brilliant chromium sparkles. Strummer says 'Alright, okay,' someone shouts something into the roar and begins a count to three.

One. In jackets butchered from Captain Scarlet, zipped and sewn with patches and stencilled with terrorist slogans, in big white creepers, drainpipe-legged and badged in that flickering hiatus they glow like gods on their pedestal.

Two. Gunslinger-stanced with a strangling grip on their weapons' necks, in red and black and blue they lock and load in a line of three, spotlit and heroic.

Three. Jones jumps, pulls his guitar up to his chin, tucks up his heels, Strummer plants one foot back and winds up his arm and Simonon turns in his toes and drops legs spread low, looking down at his fretboard. Headon taps the snare.

And then, as a diamond-bright hail of spittle rains into the light my world ignites. Total energy – raw, crashing, loud and determined, fired electrically by six, twelve, sixteen stabbed strings and blurred sticks smacking skins and all cranked up really high. I'm exultant, choking in an airless crush, in a swamp of wild strangers, I watch as Strummer stamps his foot forward, grabs the mike in another volley of spit and tells us that London's burning with boredom. Yes, and here pumping and beating, all the kids are biting in pure spite, breathless and breaking themselves apart as the white riot materialises and the Top Rank Ballroom is shaken into the fluorescent, pulsating spring of beautifully blooming punk.

Their arrogant rhetoric is irresistible to me, their posturing so proud and idealistic. The music and fashion the perfect fabric to fuel my raging need to confront my all parents' taboos, all their contradictions and confusions as well as the destroyed economy, the degenerate racist, fascist, class-ridden mess that I've realised is my seventies Britain. Whirling amongst all these apprentices of anarchy I can see that smouldering disaffection so clearly, and I love the press-fuelled fear and loathing, the farcical hysteria that is allowing me to use a few safety pins, zips and ripped T-shirts to instantly separate myself from everything and, essentially, from those classroom-cannibals who had torn me up and who I now truly hated.

I'd been bullied beyond any need to care, so with one packet of Born Blonde and some Crazy Color I bleached away the desire to satisfy anyone but myself and lit up my loneliness with a shock of blue spiked hair. With a black leather motorcycle jacket and a bag of studs I was rejecting the wrong-world and embracing an impenetrable outlook of defiance, an indestructible determination to win at all costs, in my way, on my terms.

Ironically, despite the aggro, the beatings on the street, outside gigs and at college that left me bloody-nosed and fat-lipped, punk was protecting me. I seized at any pertinent mantras mouthed by the bands, started reading about things that weren't dinosaurs, otters or Kestrels, and could hide in a place where I could concentrate my anger and stoke my blazing engines of rage. I might stumble, get battered and bruised, get locked out, get hurt, but if I could fight, work and drive myself, harder, madder, faster than anyone else then ultimately I would win. And petulantly, they hadn't wanted or understood me anyway, so now they weren't going to have me. Fuck off, fuck you, fuck it all.

September 2003

The room seemed to be buzzing with an irritable static. He was jammed into the chair gripping and re-gripping the wooden ends to the arms, sat frigid and upright and perfectly symmetrically. He'd been silent for a very long minute before he breathed, 'I forgot about the enemy.'

She consciously waited and then without looking up asked, 'Who is the enemy?'

'I am,' he replied quietly, 'I'm the enemy, my enemy. And I'm chasing myself and when I finally catch me I'm going to kill me.'

He spoke as if this scenario was a simple, obvious, everyday matter of fact so she responded in a similar vein, 'How do you stay ahead?'

A door clicked open in the hall outside but he didn't hear it. He drew a deep breath to fuel a long nasal sigh and then whispered, 'By running, by never stopping, by constantly trying to make it better, do it better. By never giving up, by always believing that I can, I must, I will.'

'Do what better?' She cast her eyes over him as he clasped the chair, becoming part of its structure, as if bracing against its imminent disintegration.

'Everything.' He nodded. 'Every single last thing. Every breath, every task, every job, everything.'

'So what went wrong?'

'He died, she left me, and I failed. And . . . ' he sighed, 'all that

tripped me up and I caught up with myself and in the ensuing chaos I couldn't get away. I had nothing left. No fuel, no energy, no direction, no perspective and ...' he paused, 'no hope.' His face was utterly expressionless.

She too now sat perfectly still, aware of her immobility. And after a respectful break said, 'And taking your own life was the solution?'

'It seemed so. And it felt right. Because this time it was only about me. It wasn't like when I'd got close to it before. That was about other people, about me planning to hurt the people who had hurt me, about passing my pain on to them. About punishment, about leaving them blaming themselves for effectively killing me, leaving them with a lifetime of impossible redemption. That was about measuring my sacrifice, my death, against their imagined suffering ...'

He paused and then said more assertively, 'And ultimately that's why it didn't happen I suppose. Because it was only imagined, their reaction wasn't guaranteed and I saw, or found, just enough to think that I'd beat them all rather than teach them all a lesson. There was just enough to get up and get going again, to get away from that part of me that seemed to consider for that moment,' he sighed, 'that failure was an option.'

He realigned, checked his composition and reset himself. A plane passed loudly overhead and they both listened to its untimely roar, and then he went on.

'But this time it was just me, no one else, they had all gone. I was alone and the decision was mine. I didn't need to justify it to anyone but me and I could justify it to myself. You see, there was no sense of wastage. I could have a happy death. Because although I hadn't had a typical happy life I'd had a life far beyond

my expectations, had riches beyond my dreams and my prize was to feel secure and satisfied that it was my time to die, that I was able to make and effect that decision. And essentially to justify it, to be able to accept it, wholly accept it.

'It will strike you as perversely paradoxical but a central part of the freedom to make that decision was actually because of all the good things that had happened – not the bad. It was like that scene at the end of Blade Runner, *when Roy, the android, realises that his remarkable life is up, when he kneels in the rain and says "I've seen things you people wouldn't believe. Attack ships on fire off the shoulder of Orion. I watched C-beams glitter in the dark near the Tannhauser gate ..." Well, I've seen those attack ships, for me they flew in fields, over oceans, through jungles, all those remarkable birds I'd studied by torchlight under the bedcovers on Brooke Bond tea-cards and never dared to dream I'd see in real life. Cock of the rock, males dancing in the half light of the Peruvian rainforest, I saw them. I saw them. And I've seen his sea beams, they were my dolphins and whales and turtles and coral reefs filled with kaleidoscopic swirls of fish ... I couldn't, wouldn't ever dare to feel that my life hasn't been fulfilled, I've had my fingers, these fingers, in the sparkle jar.'*

'I know that film,' she said, 'the whole point is that those androids, Roy, didn't want to die.'

'No he didn't want to die, he wanted more life, but when he finally realises it's over he justifies his death with the knowledge that he has been briefly but so intensely privileged. And I've had more, more than I've deserved, far more. And in that tangle of depression I couldn't see any further, I couldn't see any future, so I was comfortable with that.'

'What about your family, what about your stepdaughter? Did

you think about them, about how they would feel, about how she would cope with this on top of all the other problems in her life?'

The room was too still, like it had been sealed and solidified, like a painting, as if this moment was fixed forever.

'No. I couldn't see them. They didn't exist then. They had all gone … That's the point, that's what is always impossible for people to understand when they try to rationalise this situation, they simply can't comprehend the absolute separation. They think you can see out, but you can't. It's fucking obvious you can't otherwise you wouldn't be about to kill yourself because you'd be able to see a future – let alone your family or friends if you bloody had any left. Everyone says that suicide is selfish. How fucking stupid. Selfishness would only be apparent if you could be conscious of your actions' effect on others. But there are no fucking others, they are not there. There is nothing there … but you and a fucking, great, hopeless, vacuum.'

'So after that time when there weren't enough pills, why didn't you get some more and then go through with it?'

'Because I think that place is difficult to get back to once you're out. It's too terrifying to go straight back there. When you know that you could have done it, that you had the power, that you had made that decision, truly made that decision. That in one way you are already dead. And something does die I think … it's the life-raft, the parachute, the safety line. You know they are all gone then and the paradox is that you actually feel more alive. In the days afterwards I felt exhilarated, completely free, totally independent, like I had some sort of special superpower. But that fades and in time you begin to reach the sides, to see things, touch things again and they cling to you like sticky paper and they wrap you up and begin to weigh you down, they ground

you and make it impossible to be that alone again. So you then can't go back there. Not straight away.'

'So was that it?'

'No. Actually I did get back there again. And what was really ugly was that I knew it, I was familiar with that place. It was like suddenly finding myself in the same very empty room and whilst there's a comfort in that familiarity it's just as overpowering, as swamping. And it struck me as a bit sad, that I should have come to know this place. I felt a bit sorry for myself for having learned to recognise this degree of ... hopelessness, whatever. And just because you've got out before doesn't mean there's a way out again. You're back in that place and the situation is exactly the same as before.'

'But you didn't kill yourself?'

'The dogs,' he nodded and pinched his lips into a slight smile, 'I was on my own with the dogs. I couldn't leave them. They loved me. I hated that though, at that point I hated them for loving me. I so, so wished they weren't there, so wished ... But they loved me so I couldn't do it. They kept me alive. They did. I owe them my life. I always will. And now I love them so much. As much as I loved my Kestrel. And that's the lunacy of it isn't it? That total and utter devotion, that unconditional, complete and infallible love, that obsessive and indestructible commitment to something which cannot ever, ever fail you, lie to you, let you down, use or abuse you, that will never have an agenda other than pure love, immaculate, perfect love, is the thing that is there waiting to destroy you. Because it becomes all of you and when it's gone there is absolutely nothing left and you'll slide back into that place. Where ... nothing ... sparkles ...'

He stumbled from that tight little secret room and out into the massive unknowing and unconcerned world and grasped at the air, heaving his chest to have something to fill it. He felt punched, winded, as if a train had just smashed through him, and he literally staggered across the pavement and had to concentrate to remain standing. He stood, a warm drizzle licking his face, his eyes rolling. He stood in a state of horror, because the real and actual situation he'd just relived had finally become wholly tangible to him. He'd seen the pin-sharp proximity of his death and listened to a litany of rational explanations that had given it absolute validity. And as he'd left the clinic, the reality of that had crashed all at once, in one great tonnage, in a massive thump. All the words, all the things he'd recounted fell into him and exploded.

So he teetered there choking, gagging in the rain, gazing at trees so vividly green they appeared stencilled onto the grimace of the flinty sky. And across that sheer bruised face he thought he saw the arc of a brilliant rainbow, so beautiful and bright, so close – the red so vivid, so bloody, the violet bleeding into the crackling air, the yellow so livid. He breathed deeply, his head still struggling for clarity, and then from amongst the fangs of cloud a tiny twinkling shape appeared, a circling speck rising on pretty flaps of winnowing wings, climbing the spires of the storm, carving a helix into the updraft, tilting and turning before banking up and flaming as it washed through the rainbow's fireglow and then sailing towards him blackly through the orange and yellow and slipstreaming across the blue and violet and out over the stony vault of heaven, where it rung up thunderspun and hung hovering as bright as a star and sparkled as it filled the hole in his sky.

ACKNOWLEDGEMENTS

None of what you have read could or would have happened without the unswerving encouragement, unerring patience, enormous energy, considerable tolerance, practical assistance and incomparable educational support provided by my father. He taught me to read and bought me books, taught me to draw and let me use his prized pencils. He led me into the garden to gawp at Spitfires and then back indoors to build Airfix kits of those and all our other favourite aeroplanes. He gave me all of his weekends and drove me to museums and castles and then to all corners of the county and country in pursuit of my wildlife fantasies. Perhaps regretfully, from his perspective, he also showed me how to shoot my air rifle – the local constabulary were frequent visitors after that. Together with my mother, he allowed me to turn the house into a menagerie and the garden into a safari park. Their carpets were soiled by many species, they withstood many escapes and, for periods, they dealt with the noxious aromas that emanated from my precious museum specimens. He once helped me cook a pilot whale's head in a stew pot that he had made especially so I could 'get its skull'. He regretted that, too. Indeed, the entire neighbourhood regretted that.

My mother took me to the zoo. During a week of one summer holiday, we boarded the bus and visited the zoo three times. For me that was fabulous. I'm sure that for her it was taxing, because if my father did the history and natural history, then typically my mother did the art and literature. She took us to galleries, to all the local country mansions, to the ballet, opera and classical concerts and to the theatre. She read me poetry and got me to recite it to her, and her restless mind was always excited by some novel piece of culture we had to know about. She encouraged my writing, and for thirty years she typed all my essays, dissertations and stories, and then wrote critiques in the margins. She was never frightened of the truth, loved fluid prose and admired beauty made from words. She always said I should write a 'proper book, a serious book'. Whether this is it, I'll never know – she died a few years ago.

Despite my determination, my sister could not be led to an interest in the natural world. She baled completely aged ten when I dragged her through a dense nettle bed to see a mallard's nest. By the time the tears had dried, her interest was only in finding a profitable outlet for my enthusiasm, something she has worked steadfastly at all of our adult lives. It was her who insisted I pursue an audition for *The Really Wild Show*, no doubt hoping that, if successful, it would divert my obsessional outbursts to a wider audience and allow a little peace to finally reign over the family dining table. We don't talk about animals much these days, but we spoke a lot about this book, and for that I'm grateful.

The book was written without any publishers' initial interest. I don't, can't, like my own work, and was thus unable to make any accurate assessment of its quality or worth, so I sent it to my long-time creative mentor, Rosamund Kidman Cox. We had met in the eighties beneath the Diplodocus at the Natural History Museum, where as a judge she had been both bemused and surprised by some of my early and outrageous forays into wildlife photography. Since then, she has kindly passed comment on a litany of ideas, photos and texts, and after detailed and considered scrutiny, always pronounces her opinions with complete honesty. In this instance, I asked her, 'Should I publish this?' She pursed her lips, shook her head and replied, 'You *must* publish it.' Without my trust in her judgement, these pages would not exist. Thanks Roz.

Shortly afterwards, I met Robert Macfarlane and brazenly imposed a draft upon him. He, too, offered considerable support and advice, which was immeasurably important in many ways, not least to add buoyancy to the project and stabilise my wavering morale.

Ebury have been superb throughout. Andrew Goodfellow's immediate energy and enthusiasm for the text bowled me over, and his passion for the book has been boundless. Without it I would never have been satisfied or secure through the publication process, and I am consequently hugely indebted to him. He and Anna Mrowiec have been a complete joy to work with on every level. They have entertained and gently modified my whims and countered my insecurities with creative discussion and pragmatic deliberations. Even when I sent them a highly detailed, 15-page illustrated essay entitled 'Producing the Perfect Book Cover' and then decorated their office with a flotsam of my designs, they didn't seem to waver. But then, at our initial meeting, I'd shown Andrew some graphical representations of the narrative, so they were probably already prepared for such anomalies. Their colleagues Claire Scott, Caroline Butler and Sarah Bennie have also been an absolute pleasure to deal with.

Jo Charlesworth shot all the publicity photographs for the project. I had long admired his portraiture, and I am so pleased that he agreed to take on the task of photographing a photographic perfectionist. Not that he was too daunted, I'm sure, as his own level of critical appraisal is fabulously high, and as a result the images are both original and beautiful. Only his subject lets them down.

Together with wildlife cameraman Ian Llewellyn, Jo also shot and produced a short film to promote the book starring Ian's son, Jem, as the young Chris. It was a passion project and highlights their considerable artistic and technical skills, and I am both flattered and grateful for their interest and dedication. Nick Pitt very generously loaned us space at his Farm Studios near Bristol for both photography and filming, and the incomparable Lloyd and Rose Buck allowed us to use Ashley, their wonderful male Kestrel.

Charlotte had to negotiate the 'creative aura' (that's me existing in a totally unresponsive state and not being able to think about anything else) and the innumerable frustrations (that's me ranting about all sorts of things I can only ever rant to her about) from which the words were wrung, alongside her long-term management of my personality traits. No mean feat. Thanks Fos.

And lastly there's the Itch and the Scratch, without whom nothing would mean anything at all. Woof, woof.